Adaptation to Climate Change and Development

Adaptation to Climate Change and Development

Editor: Daisy Mathews

MURPHY & MOORE
www.murphy-moorepublishing.com

www.murphy-moorepublishing.com

MURPHY & MOORE

Cataloging-in-publication Data

Adaptation to climate change and development / edited by Daisy Mathews.
p. cm.
Includes bibliographical references and index.
ISBN 978-1-63987-691-4
1. Climatic changes. 2. Sustainable development. 3. Economic development--Environmental aspects.
4. Adaptation (Biology). 5. Climatology. I. Mathews, Daisy.
QC903 .A333 2023
363.738 74--dc23

Murphy & Moore Publishing
1 Rockefeller Plaza,
New York City,
NY 10020, USA

ISBN 978-1-63987-691-4

TABLE OF CONTENTS

Preface

The main aim of this book is to educate learners and enhance their research focus by presenting diverse topics covering this vast field. This is an advanced book which compiles significant studies by distinguished experts in the area of analysis. This book addresses successive solutions to the challenges arising in the area of application, along with it; the book provides scope for future developments.

Contemporary climate change encompasses global warming and its effects on Earth's weather patterns. Climate change adaptation refers to the process of adjusting to existing or anticipated effects of climate change. It has the ability to significantly decrease many of the negative effects of climate change and increase the favorable effects. Climate adaptation entails behavioral changes in individuals and constructing better infrastructure to manage altering weather patterns. Adaptive capacity is also inextricably linked to economic and social development. The solutions for climate change adaptation usually differ for different communities, organizations, countries, and regions. Adaptation measures may include building flood defenses, switching to drought-resistant crops, redesigning communication systems, setting up early warning systems for cyclones, and reconsidering business operations and government policies. Climate change can be viewed in two ways, as a problem of development, and as a problem for development. The policies and practices with respect to development play an important role in addressing climate change. This book strives to provide a fair idea and help develop a better understanding of the latest advances in the study of climate change adaptation and development. It is a resource guide for experts as well as students.

It was a great honour to edit this book, though there were challenges, as it involved a lot of communication and networking between me and the editorial team. However, the end result was this all-inclusive book covering diverse themes in the field.

Finally, it is important to acknowledge the efforts of the contributors for their excellent chapters, through which a wide variety of issues have been addressed. I would also like to thank my colleagues for their valuable feedback during the making of this book.

Editor

1

Drivers of Level of Adaptation to Climate Change in Smallholder Farming Systems in Southern Africa: A Multilevel Modeling Approach

Byron Zamasiya, Kefasi Nyikahadzoi and Billy Billiard Mukamuri

Contents

Abstract

Climate change is a major development challenge affecting developing countries that rely on rain-fed agricultural production for food and income. Smallholder farmers in these countries are using multiple adaptation practices to manage the effects of climate change. This chapter examines household and community-level factors that influence smallholder farmers' level of adaptation to climate change in the Hwedza District in Zimbabwe. Data for this study were collected from 400 randomly selected smallholder farmers, using a structured questionnaire, focus group discussions, and key informant interviews. The study used a multilevel modeling approach to examine the factors that influence smallholder farmers level of adaptation to climate change. Results from the study show that smallholder farmers' level of adaptation to climate change is conditioned by access to extension services, access to remittances, family labor, household education

B. Zamasiya (✉) · K. Nyikahadzoi · B. B. Mukamuri
Centre for Applied Social Sciences, University of Zimbabwe, Harare, Zimbabwe
e-mail: bzamasiya@gmail.com; knyika@gmail.com; bmukamuri@gmail.com

(household level factors), and linking capital (community-level factor). This chapter therefore concludes that smallholder farmers that have higher levels of adaptation to climate change are those that are well linked to external organizations and have access to agricultural extension services. The chapter recommends that adaptation to climate change can be enhanced by improving access to agricultural extension services and promoting linkages with external organizations that provide information on agricultural adaptation practices.

Keywords

Level of adaptation · Climate change · Smallholder farmers · Hierarchical linear modeling · Linking capital

Introduction and Background

Climate change is a major development challenge facing developing countries particularly those in Southern Africa (Ali and Erenstein 2017; IPCC 2014). These countries are battling to address multiple stressors that include a high incidence of poverty, chronic food insecurity, malnutrition, HIV, lack of disaster preparedness, political upheavals, and macroeconomic instability (Adego et al. 2018). In this region, rain-fed agriculture which accounts for 95% of agricultural production provides livelihoods to over 70% of the population (Zamasiya and Nyikahadzoi 2018; Asfaw et al. 2016). Overreliance on climate-sensitive agricultural systems exacerbates the vulnerability of countries in Southern Africa such as Zimbabwe (Rurinda et al. 2013). Rainfall is the major determinant of agricultural production. Reports show that the mean annual rainfall has already declined by over 5% and forecasted to further decrease by 10%. The IPCC's 2018 report notes that with business as usual approach to greenhouse emissions, temperatures will increase by over 1.5 °C. The consequences of such an increase will be catastrophic to resource-constrained smallholder farmers in Southern Africa particularly those with low adaptive capacity. This development will plunge developing countries into dire food insecurity and chronic malnutrition. These conditions require that smallholder farmers who are the major food producers adapt to climate change (Abid et al. 2015; Mabe et al. 2012). Adaptation, in this case, is the use of farming practices that can reduce the impact of climate change on smallholder farmers' food security.

Scholars observe that adaptation reduces the impacts of climate change on food security systems (Adego et al. 2018; Shisanya and Mafongoya 2016; Di Falco et al. 2011). Some of the adaptation practices that smallholder farmers are to manage the impacts of climate change include crop diversification, drought-tolerant varieties, income diversification, staggering of planting dates, crop mixing, and soil and water conservation (Adego et al. 2018; Shisanya and Mafongoya 2016; Nyikahadzoi et al. 2017). Studies show that some smallholder farmers use more than one adaptation practices. In such circumstances, such smallholder farmers are likely to have better food security than those that use solitary adaptation practices. Although adaptation is a viable response to climate change, smallholder farmers face challenges in implementing the adaptation practices. These challenges include lack of access to climate information, lack of access to agricultural credit, lack of access to an extension to adequately prepare for seasons, lack of knowledge on adaptation practices, and lack of labor resources (Nyikahadzoi et al. 2017; Salau et al. 2012; Di Falco et al. 2011). Since adaptation differs by contexts, some of the adaptation practices may not be applicable in other regions.

There are very limited studies (Ali and Erenstein 2017; Below et al. 2012) on drivers of the level of adaptation to climate change by smallholders. While the limited studies that are there provide important insights on determinants of the level of adaptation among smallholder farmers, their major challenge is that they used simplistic modeling techniques. These techniques do not recognize the nesting of data in groups and may, therefore, lead to biased results and poor policy targeting. This chapter recognizes that since farmers are nested in villages and villages are nested in wards, the actions of individual smallholder farmers are a function of household socioeconomic level and community-level variables. This is because smallholder farmers live in an open environment in which they interact among themselves, interact with people from other wards, and interact with external organizations. The levels of interaction vary from village to village, and this necessitates the use of multilevel modeling techniques (Frankenberger et al. 2013). The chapter proposes the use of the hierarchical linear modeling technique to examine the drivers of the level of adaptation to climate change among smallholders' farmers in the Hwedza District of Zimbabwe. This model addresses the weaknesses of confounding that are inherent in models used in previous studies.

In some of the studies on drivers of the level of adaptation to climate change, social capital was used as an explanatory variable. The definition used for social capital is restricted to farmer-farmer interactions at a local level such as within a village. This definition does not capture vertical interactions between communities and organizations that are external to the community. In this study, a deliberate attempt is made to distinguish between social capitals. The chapter adopts the categorization of social capital proposed by Aldrich (2012) and Frankenberger et al. (2013) which provides a clear distinction between bonding capital, bridging capital and linking capital. In this study, our concern is on linking social capital. It refers to the linkages between communities and organizations that are external to it. Scholars argue that communities that have better resilience are those that have higher linking social capital compared to bonding capital (Aldrich 2012; Taruvinga et al. 2017). This chapter measures linking capital as a village-level variable and use it as a level 2 covariate in the hierarchical linear regression model.

Given the inevitability of climate change, adequately preparing for this eventuality requires policymakers to understand the drivers of the level of current adaptation practices to climate change by smallholders. The purpose of this chapter is to improve policymakers' understanding of the determinants of smallholder farmers' level of adaptation to climate change. The issue of determinants of the level of adaptation to climate change among smallholder farmers is poorly understood and measured. In this study, the level of adaptation to climate change is the number of strategies that are used by a smallholder farmer in response to climate change. This chapter, therefore, seeks to provide empirical evidence on what drives the level of adaptation. This information is critical for identifying and setting policy targets for enhancing adaptation. Without that, it is difficult for policymakers to enhance current and future adaptation by smallholder farmers.

Data Gathering and Analytical Framework

This study was conducted in Hwedza District in Mashonaland East Province of Zimbabwe is a "climate hotspot." Zimbabwe is divided into five agro-ecological regions known as natural regions (NR). This classification is based on the soil quality, rainfall regime, and vegetation distribution. The quality of the land resource declines from natural region (NR) I through to NR V. Hwedza District is located in natural region II with an annual rainfall of 650–800 mm and a mean temperature of 29 °C (Rurinda et al. 2013). The predominant soil type is coarse sandy soils derived from granitic rocks (Rurinda et al. 2014). Average household farm sizes range between 2 and 5 ha per household (Rurinda et al. 2014). Hwedza District is a communal area dominated by subsistence maize production for food security. The subsistence farmers also grow groundnuts, finger millet, and cowpeas on small plots (Zamasiya et al. 2018). Crop production is complemented with livestock rearing mostly cattle, goats, and indigenous chickens. During poor cropping seasons, smallholder farmers in the district rely on non-timber forest products for food (Woittiez et al. 2013) and food aid from nongovernmental organizations (NGO). Due to climate change, the onset and distribution of rainfall in the Hwedza District are now very unpredictable. The district experiences frequent midseason droughts with good rainfall years increasingly becoming fewer over the past 20 years (Rurinda et al. 2014). Smallholder farmers in this district use different farming practices to manage the effects of climate change. These practices include crop diversification, staggering of planting dates, soil and water conservation practices, drought-tolerant varieties, integrated soil fertility management, and crop mixing (Zamasiya et al. 2017). Public and private research organizations and nongovernmental organizations were promoting a basket of ex ante climate adaptation practices to improve smallholder farmers' preparedness to droughts and increased rainfall variability.

The study used a multistage sampling technique to select the study site and the survey respondents. The target population in this study are smallholder farmers practicing crop and/or livestock farming in Hwedza District in Zimbabwe. In stage one, purposive sampling was used to select the Hwedza District in Mashonaland East Province of Zimbabwe. This study site was selected on the basis that it has a high agro-ecological potential and has experienced an increase in extreme climatic events since the year 2000. In stage two, random sampling was used to select Ushe and Dendenyore wards from a total of 22 wards in the district. In stage 3, simple random sampling was used to select 10 villages per ward. Dendenyore ward has 30 villages, and the Ushe ward has 22 villages. In the last stage, the study used simple random sampling to select 20 households in each village. In total, 400 smallholder farmers were interviewed during the survey in February 2015.

Quantitative and qualitative data used in this study was solicited from smallholder farmers using a structured questionnaire, focus group discussions (FGDs) and key informant interviews. The structured questionnaire collected data on demographic information, strategies used by smallholder farmers to respond to climate change, access to extension, remittances, and challenges experienced by farmers in adapting to climate change. The key informant interviews were conducted with ward level agricultural extension officers (AEOs) and district level extension officers. In total, eight AEOs participated in the key informant interviews. The study also collected data from knowledgeable smallholder farmers. A total of six farmers participated in the key informant interviews. Knowledgeable farmers are smallholders who have at least 30 years of conducting farming activities in the Hwedza district. Purposive sampling was used to select participants for the key informant interviews. Focus group discussions were conducted with smallholder farmers in each village. Each FGD had 12 smallholder farmers (6 males and 6 females). The purpose of FGDs in each village was to get an average rating of the village's perception of the level of linking the capital, the farming practices used by smallholder farmers, and the challenges that they are experiencing in adapting.

Quantitative data analysis was done in STATA Version 15. This study used a multilevel modeling technique that is known as the hierarchical linear modeling (HLM) to establish the determinants of farmers' level of adaptation to climate change. An HLM technique is a random coefficient modeling approach that can be used to analyze data that is nested within groups (Huta 2014). Such models are used to examine how individual-level relationships vary as a function of group characteristics (Hox et al. 2018). The HLM estimation procedures are capable of analyzing nested data (Huta 2014). The results from this analysis are presented in tables as coefficients. This study used a level 2 HLM approach comprising household-level variables for individual farmers and group level variables. Qualitative data collected from interviews and focus group discussions and observations were analyzed thematically.

Multilevel Estimation of Determinants of Level of Adaptation

This chapter presents the results from the determinants of the level of adaptation to climate change among smallholder farmers in Zimbabwe.

The first stage is to run the random effects model. The purpose of doing this is to understand the portion of variance that is due to cross-village interactions compared to individual household interactions. This model can be written as follows:

Random Effects

Results in Table 1 show that there are 400 level 1 units (number of observations) in the analysis and 20 level 2 units (number of villages/groups). The cluster size varies from 14 to 28 with a mean cluster size of 20. The log-likelihood ratio is −797, 1447. The grand mean across the villages is 1.5066 with a standard error of 0.1640. The z

Table 1 Estimation results for a random effects model

LOA	Coef.	Std. Err.	Z	P > z	[95% Conf.	Interval]
_cons	1.5066	0.1640	9.1900	0.0000	1.1851	1.8281
Random effects parameters		Estimate	Std. Err.	[95% Conf. Interval]		
village: Identity						
var(_cons)		0.3868	0.1723	0.1615	0.9263	
var(Residual)		2.9565	0.2146	2.5644	3.4085	

LR test versus linear model: chibar2 (01) = 22.5800 Prob ≥ chibar2 = 0.0000

Table 2 Estimation results for a random intercept model

Adaptation (Adj)	Coef.	Std. Err.	z	P > z	[95% Conf.	Interval]
Visitextn	1.1785	0.2022	5.8300	0.0000***	0.7822	1.5748
FamilyRemit	0.7180	0.2262	3.1700	0.0020***	0.2747	1.1612
Farmlabor	0.1669	0.0433	3.8500	0.0000***	0.0819	0.2518
Hheducation	0.0533	0.0316	1.6900	0.0920*	−0.0087	0.1152
_cons	0.3953	0.2588	1.5300	0.1270	−0.1119	0.9025
Random effects parameters		**Estimate**	**Std. Err.**	**[95% Conf. Interval]**		
village: Identity						
var(_cons)		0.2390	0.1193	0.0898	0.6360	
var(Residual)		2.6064	0.1892	2.2608	3.0049	

Wald chi2(4) =59.92; Prob > chi2 = 0.0000; LR test versus linear model: chibar2(01) = 13.61 Prob ≥ chibar2 = 0.0001

statistic is 9.1900, and the p-value is 0.000. The reported LR test statistic is 22.5800 with a p-value of 0.000; we, therefore, reject the null hypothesis that tao is equal to zero. It, therefore, means that there is intervillage variation in the adaptation by smallholder farmers. Based on the output, the interclass correlation is 11.57%. This means that 11.57% of the variation in adaptation among smallholder farmers in Hwedza district is due to differences across villages and 88.43% is due to individual differences across farmers.

The random intercept model involves the addition of level 1 covariates to the model, but it assumes fixed effects. In this stage, the intercept is allowed to vary across villages to accommodate intervillage variations. The model can be expressed as follows:

Table 2 shows the estimation results of the random intercept model. The results show that all the independent variables have positive effects on adaptation. They are all statistically significant with Visitextn (p <1%), FamilyRemit (p <1%), Farmlabor (p <1%), and Hheducation (p <10%). The Wald test shows that W = 59.92 and p-value =0.0000. Comparing the fit of the random intercept model and the LR = 13.61 and a p-value of 0.0001. Therefore, the null hypothesis that the intercept is the same across all villages is rejected.

Hierarchical Linear Regression Model with Random Intercept and Level 2 Covariates

This model seeks to account for the variation in intercepts across villages. This is done by adding a level two covariate.

Table 3 shows the results of the hierarchical linear regression model with random incept and a village-level covariate. The results show that Level 1 covariates are all significant. The level 2 covariate (linking social capital) is also very significant ($p < 1\%$). The Wald test statistic W = 100.55 has a p-value of 0.000. The results show that the second level variable linking social capital improves the level of adaptation for smallholder farmers.

This chapter presents the discussion from the study. The discussion is based on theoretical and empirical evidence to recommend how government and organizations external to the community can help in improving adaptation to climate change by smallholder farmers. Based on this study, the variables that can be influenced to promote adaptation are access to agricultural extension services, access to remittances, family labor, level of education, and linking social capital.

The coefficient of access to agricultural extension services ($p < 1\%$) is statistically significant and positively influences smallholder farmers' level of adaptation to climate change. Extension officers play a pivotal role in the dissemination of agricultural adaptation technologies (Zamasiya and Nyikahadzoi 2018; Zamasiya et al. 2014). Through field-based demonstrations and awareness meetings, extension officers help in disseminating information to smallholder farmers on adaptation practices that they can use to manage the impacts of climate change on food security (Zamasiya et al. 2017; Ali and Erenstein 2017). Smallholder farmers who have access to AEOs are more likely to know different adaptation practices. As such they are more likely to have higher levels of adaptation to climate change than their counterparts. However, key informants highlighted that the major challenge facing AEOs is Zimbabwe is lack of motorcycles for improving mobility and contact with

Table 3 Estimation results for HLM with random intercept and level 2 covariates

LOA	Coef.	Std. Err.	Z	P > z	[95% Conf. Interval]	
Extension	1.0950	0.1995	5.4900	0.0000	0.7040	1.4859
FamilyRemit	0.6645	0.2234	2.9700	0.0030	0.2267	1.1022
Farmlabor	0.1574	0.0428	3.6700	0.0000	0.0735	0.2414
Hheducation	0.0554	0.0312	1.7800	0.0760	−0.0057	0.1165
Linking	2.1973	0.4137	5.3100	0.0000	1.3865	3.0081
_cons	−0.4074	0.2734	−1.4900	0.1360	−0.9433	0.1286
Random effects parameters		Estimate	Std. Err.	[95% Conf. Interval]		
village: Identity						
var(_cons)		0.0216	0.0493	0.0002	1.8808	
var(Residual)		2.6050	0.1890	2.2597	3.0031	

Wald chi2(5) = 100.55; Prob > chi2 = 0.0000; LR test versus linear model: chibar2(01) = 0.23 Prob ≥ chibar2 = 0.3150 smallholder farmers. The wards in Hwedza District are so big that it becomes difficult for the AEOs to effectively contact farmers if they are not motorized. Other AEOs highlighted that they have motorcycles but have no access to fuel. This, therefore, makes it difficult for them to dispatch their roles efficiently. During FGDs, smallholder farmers highlighted that the challenge with AEOs is that climate change is a new phenomenon to them. They were advising farmers based on their lived realities. These results corroborate with the findings of Zamasiya et al. (2018), and Ali and Erenstein (2017).

Results show that the coefficient of access to remittances ($p < 1\%$) is very statistically significant and positively improves a smallholder farmer's level of adaptation to climate change. Remittances improve the number of financial resources that are available to a household (Nyikahadzoi et al. 2013) for purposes of financing transaction costs associated with adaptation to climate change. These costs could include the cost of improved seed, inorganic fertilizers, and herbicides. Zamasiya et al. (2018) note that smallholder farmers with access to remittances are more likely to adapt than their counterparts. They further highlight that some smallholder farmers rely on disposing of small livestock to finance adaptation costs. This is the case in economies such as Zimbabwe where most smallholder farmers have no access to formal banking systems (Nyikahadzoi et al. 2012). With improved access to financial resources, it is more likely that the smallholder farmers will improve their levels of adaptation. During FGDs, smallholder farmers highlighted that their major challenge is lack of agricultural credit. They mentioned that the lack of access to financial resources limited their ability to implement several adaptation practices. For instance, some farmers highlighted that they lacked financial resources to purchase improved seeds, inorganic fertilizers, and herbicides, and this made it difficult to implement several adaptation practices. Some of the smallholder farmers ended up planting recycled maize seeds. Key informants highlighted that when smallholder farmers used recycled seed, they used the wrong seed rate for maize crop. The key informants also highlighted that in most cases, agro-dealers sell only one variety of maize seeds. The smallholder farmers just end up buying the available seed. These results corroborate with the findings of Ali and Erenstein (2017).

Study findings show that the coefficient of family labor positive and statistically significant ($p < 1\%$) influences a smallholder farmer's level of adaptation to climate change. Labor is a major constraint to adaptation among smallholder farmers (Deressa et al. 2010). The family offers technical and manual skills that are crucial for executing agronomic practices on time. Smallholder farming households that have large pools of labor can spread labor resources across different adaptation practices. Such households can utilize labor-intensive practices such as crop diversification and soil and water conservation practices (Zamasiya et al. 2018). As such, households with higher labor resources are more likely to have higher levels of adaptation than their counterparts. These results corroborate with the findings of Ali and Erenstein (2017).

Research findings show that the coefficient of level of education ($p < 10\%$) is marginally statistically significant and positively influences a smallholder farmers' level of adaptation to climate change. Zamasiya et al. (2018) note that education improves a smallholder farmer's ability to receive, decode, and use the information on adaptation practices. These results suggest that smallholder farmers who are educated are more likely to be aware of climate change, the adaptation practices that they can use, and the benefits of adaptation. It is therefore likely that such smallholder farmers are more likely to have higher levels of adaptation to climate change than their counterparts. During FGDs, it was observed that most educated smallholder farmers were aware of adaptation practices and were proactively seeking more information on the practices from extension officers. Key informants highlighted that educated farmers were willing to experiment with new adaptation practices than uneducated smallholder farmers. Key informants highlighted that the lack of education among smallholder farmers is a major constraint to adaptation. Results corroborate with the findings of Below et al. (2012) and Ali and Erenstein (2017).

Study findings show that the coefficient of linking capital (p <1%) is very statistically significant and positively influences the level of adaptation to climate change. Linking social capital is an important driver of the level of adaptation to§ climate change by smallholder farmers. As observed by Aldrich (2012), linking capital improves adaptation through the provision of information that is otherwise unavailable. Through vertical relations with organizations external to the community, linking social capital improves access to information that is very important for enhancing adaptation to climate change (Taruvinga et al. 2017; Babaei et al. 2012). Information that communities may access through linking social capital includes improved varieties, yield robustness of such varieties, baskets of adaptation technologies, research results, and opportunities for exchange visits. In the study site, the communities are linked to research institutions that conduct farmer-managed and researcher-managed experiments. The smallholder farmers participate in field days where they get to interact with staff from organizations that play a key role in the agricultural value chain. Through these platforms, smallholder farmers have access to valuable agronomic information that they use for enhancing their adaptation to climate change. During FGDs, smallholder farmers highlighted that external organizations played a pivotal role in the dissemination of information on adaptation practices. The smallholder farmers noted that they were now aware of new crop varieties, their performance under different management regimes, and had acquired invaluable knowledge on different adaptation practices. Key informants highlighted that financial resources were scarce for the Agritex Department. As such, access to linking capital is associated with higher levels of adaptation to climate change.

This chapter sought to understand the determinants of the level of adaptation among smallholder farmers in Southern Africa using the example of the Hwedza district in Zimbabwe. The study used the multilevel modeling approach. Results show that level of education, access to agricultural extension, family labor, and access to remittances positively influence the level of adaptation. At a community level, adaptation is influenced by access to linking social capital. We, therefore, conclude that smallholder farmers that have a high level of adaptation to climate change are those that are well linked to external organizations and have access to extension services and financial resources.

Promote Learning Through Linkages with Public and Private Extension Services

Based on the results from this study, the chapter recommends that to improve smallholder farmers' level of adaptation to climate change, there is a need to enhance access to agricultural extension services and NGOs or research organizations. Because the mobility of extension staff is a recurrent problem in most areas, the public extension should establish linkages with other extension service providers in their area such as NGOs and private companies so that they can combine their programs. This will ease the mobility challenges. Further, government can also boost extension service delivery through timely recruitment, periodic training of the agents, and provision of adequate logistical support. The role of organizations that are external to the smallholder farming communities cannot be underestimated. These institutions provide a feedback loop between grassroots and policymakers. They gather critical information on climate change technologies and disseminate it to smallholder farmers. For instance, NGOs and research organizations have pioneered agricultural technologies and trainings that have since been incorporated into the public extension system. To promote development of linking capital, policymakers need to adopt participatory and inclusive decision-making processes in their work with communities, encourage smallholder farmers to work in groups, promote look and learn tours to other communities to challenge, inspire, and educate communities. This will improve access to critical information on agricultural technologies that are useful for enhancing smallholder adaptation to climate change.

Acknowledgments The authors would like to thank the International Foundation for Science who provided funding for this research through Grant No. S/5158-1.

References

Abid M, Scheffran J, Schneider U, Ashfaq (2015) Farmers' perceptions of and adaptation strategies to climate change and their determinants: the case of Punjab province, Pakistan. Earth Syst Dyn 6:225

Adego T, Simane B, Woldie G (2018) The impact of adaptation practices on crop productivity in Northwest Ethiopia: an endogenous switching estimation. Dev Stud Res 6(1):129–141

Aldrich DP (2012) Building resilience: social capital in post-disaster recovery. University of Chicago Press, Chicago

Ali A, Erenstein O (2017) Assessing farmer use of climate change adaptation practices and impacts on food security and poverty in Pakistan. Clim Risk Manag 16:183–194

Asfaw S, Di Battista F, Lipper L (2016) Agricultural technology adoption under climate change in the Sahel: micro-evidence from Niger. J Afr Econ 25(5):637–669

Babaei H, Gill S, Ahmad N (2012) Bonding, bridging and linking social capital and psychological empowerment among squatter settlements in Tehran, Iran. J Basic Appl Sci Res 2(2):109–115

Below TB, Mutabazi KD, Kirschke D, Franke C, Sieber S, Siebert R, Tscherning K (2012) Can farmers' adaptation to climate change be explained by socio-economic household-level variables? Glob Environ Chang 22(1):223–235

Deressa T, Hassan R, Ringler C (2010) Perception and adaptation to climate change: the case of farmers in the Nile Basin of Ethiopia. J Agric Sci 149:23–31

Di Falco S, Veronesi M, Yesuf M (2011) Does adaptation to climate change provide food security? A micro-perspective from Ethiopia. Am J Agr Econ 93(3):829–846

Frankenberger T, Mueller M, Spangler T, Alexander S (2013) Community resilience: conceptual framework and measurement feed the future learning agenda. Westat: United States Agency for International Development (USAID), Rockville

Hox J, Moerbeek M, van de Schoot R (2018) The basic two-level regression model. In: Multilevel analysis. Techniques and applications. Routledge, New York

Huta V (2014) When to use hierarchical linear modeling. Tutor Quant Methods Psychol 10 (1):13–28

IPCC (2014) Climate Change 2014: Synthesis Report. Contribution of Working Groups I, II and III to the Fifth Assessment Report of the Intergovernmental Panel on Climate Change [Core Writing Team, RK Pachauri and LA Meyer (eds)]. IPCC, Geneva, Switzerland, 151 pp

Mabe F, Sarpong D, Osei-Asare Y (2012) Adaptive capacities of farmers to climate change adaption strategies and their effects on rice production in the Northern Region of Ghana. Russ J Agric Socio-Econ Sci 11(11):9–17

Nyikahadzoi K, Siziba S, Mango N, Mapfumo P, Adekunle A, Oluwole F (2012) Creating food self-reliance among the smallholder farmers of eastern Zimbabwe: exploring the role of integrated agricultural research for development. J Food Secur 4:647–656

Nyikahadzoi K, Zamasiya B, Muchinako G, Dziro C (2013) Enhancing social support system for improving food security among the elderly headed household in communal areas of Zimbabwe. J Food Res 2(3):46–54

Nyikahadzoi K, Mombo O, Zamasiya B, Warinda P (2017) Enhancing household food security under changing climatic conditions: a case study of Gokwe North and Hurungwe Districts in Zimbabwe. J Agric Food Inf 18(2):96–109

Rurinda J, Mapfumo P, van Wijk MT, Mtambanengwe F, Rufino MC, Chikowo R, Giller KE (2013) Managing soil fertility to adapt to rainfall variability in smallholder cropping systems in Zimbabwe. Field Crop Res 154:211–225

Rurinda J, Mapfumo P, van Wijk M, Mtambanengwe F, Rufino M, Chikowo R, Giller K (2014) Sources of vulnerability to a variable and changing climate among smallholder households in Zimbabwe: a participatory analysis. Clim Risk Manag 3:65–78

Salau ES, Onuk EG, Ibrahim A (2012) A micro-econometric analysis of adapting portfolios to climate change: adoption of agricultural systems in Latin America. J Agric Ext 16(2):199–211

Shisanya S, Mafongoya P (2016) Adaptation to climate change and the impacts on household food security among rural farmers in uMzinyathi District of Kwazulu-Natal, South Africa. Food Secur 8:597–608

Taruvinga B, Ndou P, Hlerema I, Maraganedzha T, Du Plooy P, Venter S (2017) Fostering linking social capital for successful agricultural development projects in South Africa. Agrekon 56(1):28–39

Woittiez LS, Rufino MC, Giller KE, Mapfumo P (2013) The use of woodland products to cope with climate variability in communal areas in Zimbabwe. Ecol Soc 18(4):24

Zamasiya B, Nyikahadzoi K (2018) Intensification of soybean cultivation by smallholders: opportunities and constraints. In: Nguyen H (ed) Special issue on cultivation of soybeans, vol 1. Burleigh Dodds Science Publication, Cambridge, UK, pp 305–316

Zamasiya B, Mango N, Nyikahadzoi K, Siziba S (2014) Determinants of soybean market participation by smallholder farmers in Zimbabwe. J Dev Agric Econ 6(2):49–58

Zamasiya B, Nyikahadzoi K, Mukamuri BB (2017) Factors influencing smallholder farmers' behavioral intention towards adaptation to climate change in transitional climatic zones: a case study of Hwedza District in Zimbabwe. J Environ Manag 198:233–239

Zamasiya B, Nyikahadzoi K, Mukamuri B (2018) Drivers of adaptation to climate change by smallholder farmers residing in Zimbabwe's communal areas: a case study of Hwedza District in Zimbabwe. Zambezia J Human 45(ii):75–92

2

Rethinking Climate-Smart Agriculture Adoption for Resilience-Building Among Smallholder Farmers: Gender-Sensitive Adoption Framework

Sizwile Khoza, Dewald van Niekerk and Livhuwani Nemakonde

Contents

S. Khoza (✉) · D. van Niekerk · L. Nemakonde
Unit for Environmental Sciences and Management, African Centre for Disaster Studies, North-West University, Potchefstroom, South Africa
e-mail: sizwilenk@gmail.com; dewaldvn@acds.co.za; livhuwani.nemakonde@acds.co.za

Abstract

This study identifies the need for holistic understanding of gender-differentiated climate-smart agriculture (CSA) adoption by smallholder farmers who are at the frontline of climate-related hazards and disasters in Africa. CSA adoption is predominantly informed by a parochial linear approach to farmers' decision-making processes. Resilience-building and adaptation, which forms the second pillar of CSA and can enhance understanding of the CSA adoption nuances at farmer level, often receives less attention in adoption investigations. To appreciate CSA adoption from a resilience perspective, this study focused on resilience-building based on the interlinkage between CSA and disaster risk reduction and applied a resilience perspective in a gendered approach to CSA adoption by smallholder farmers. Through primary data collected in an exploratory sequential mixed method design, the study presents a proposed normative gender-sensitive CSA adoption framework to guide CSA implementation strategies and policies. The framework is anchored in resilience thinking, and some of its key components include gender-sensitive CSA technology development, risk-informed decision-making by heterogeneous smallholder farmers, gender-sensitive enabling factors, resilience strategies, gender equitable and equal ownership, and control of and access to resilience capitals. The proposed framework can be used to improve CSA adoption by smallholder farmers by addressing gendered vulnerability and inequality that influence low adoption.

Keywords

Climate-smart agriculture · Disaster risk reduction · Gender · Adoption · Resilience · Framework

Introduction

Climate change threatens the achievement of sustainable development, undeniably presenting complex developmental challenges in less developed regions. The impetus is to find solutions to the dilemmas that disasters related to climate change present, especially in regions such as Southern Africa, which are categorized as climate change hotspots (Müller et al. 2014). In recent agricultural seasons, Southern Africa has faced devastating and unprecedented climate change-related disasters, often resulting in Member States declaring a state of disaster. Some climate-related disasters affect only one country, such as the floods in Malawi in the 2014–2015 agricultural season that caused the country to declare a state of disaster due to floods (Murray et al. 2016). Others transcend national boundaries, for example, the El Niño-Southern Oscillation-induced drought of 2015–2016 that caused Eswatini, Lesotho, Malawi, Namibia, and Zimbabwe to declare state of drought emergency (Nhamo et al. 2019). Other examples include the fall armyworm *Spodoptera frugiperda* infestation of 2017–2018 and cyclones Idai and Kenneth in the

2018–2019 season. The result of these disasters is that the food security, poverty alleviation, and sustainable development ambitions of individual countries and the affected regions are hampered (Lipper et al. 2018). One of the chief concerns is the effect of climate-related disasters on smallholder farming, which in most developing countries is estimated to constitute at least 70% of the population, with the agriculture sector contributing at least a third of gross domestic product (Diao et al. 2010).

Consequently, current development discourse in Africa is concerned with exploring resilience-building strategies for smallholder farming households so that they may become resilient in the face of climate-related disasters (Speranza et al. 2014; Lipper et al. 2018). With each climate-related disaster, there is a growing need to change from conventional agricultural farming toward new, unfamiliar, and uncommon farming technologies perceived to contribute toward resilience-building. It is for this reason that climate-smart agriculture (CSA) has gained prominence as a possible panacea to the developmental challenges presented by climate change, specifically in relation to smallholder farming in Africa (Arslan et al. 2018). CSA recognizes that climate change amplifies developmental challenges, hence its conceptualization based on three pillars, viz., (1) improved food and agricultural productivity, (2) resilience-building and adaptation, and (3) mitigation through the reduction of greenhouse gas emissions from agricultural activities (Asfaw et al. 2015; Chandra et al. 2017a). CSA is a livelihood-oriented integration of the triple wins of sustainable intensification, resilience-building, and climate mitigation (Taylor 2018). The adoption of CSA technologies and strategies provides one option for resilience-building. Consequently, there is growing focus on adoption of CSA technologies by smallholder farmers (Kpadonou et al. 2017; Nyasimi et al. 2017), although there are still some shortcomings in the understanding of CSA adoption. This chapter seeks to offer an in-depth understanding of the tensions between gender inequality and CSA adoption and of the existing limitations to resilience-building through CSA. The research departed from a resilience perspective applied on the CSA pillar that seeks to build resilience and promote the adaptation of smallholder farmers. It was conducted in two regions sharing almost similar disaster profiles, namely, Malawi and Zambia.

Findings show that low CSA adoption can be attributed to gender disparities in ownership of resilience capitals, inadequate provision for equal participation of smallholder farmers in CSA technology development, a lack of diverse CSA options that farmers could adopt, and the failure to sustain household food security, income generation, and improved quality of life through CSA. This chapter accentuates that increasing climate risk compels an exploration of measures to address these shortcomings of CSA. Furthermore, the paper emphasizes heightened need to pursue alternative gender-sensitive pathways that may help address gender disparities. Such disparities in smallholder farming societies continue to be a barrier not only to CSA adoption but also to resilience-building in the face of climate change. In pursuit of alternative approaches to addressing barriers to CSA adoption and increasing its uptake by smallholder farmers, this chapter presents a normative gender-sensitive CSA adoption framework that can be adapted and used in developing regions, ultimately contributing to resilience-building.

Criticism Directed at CSA

The conceptualization of CSA envisions that humanity could address some of its developmental challenges, such as the negative effects of climate change, population growth, and the corresponding increases in food demand, poverty, and sustainable development (Williams et al. 2015). Unfortunately, despite its positive attributes, CSA has also been met with some skepticism that cannot be ignored in a gendered approach to CSA adoption. The main criticism is insufficient consideration of power relations and inequalities (Chandra et al. 2017b). While Taylor (2018) considers power disparities at a global level between countries of the North and the South, this study considers these aspects at a farmer level. Further dissentions over CSA emanate from its failure to promote participation of local communities, with technologies and research dominantly unidirectional and top-down (Chandra et al. 2017a). Other scholars caution that when CSA fails to pay attention to social issues, its implementation may actually magnify preexisting social imbalances such as gender inequality (Murray et al. 2016; Collins 2017). When considered within the context of the pivotal role women play in smallholder farming, current CSA scholarship has insufficiencies when it comes to the appreciation of the gender dimensions in the CSA adoption decision-making process. Yet, for many African societies, the gender composition in the smallholder farming sector validates the relevance of gender as an investigative element.

Previous work by Khoza et al. (2019) shows that underlying gender inequality, patriarchy, and other social imbalances manifest as gender-differentiated, sociocultural, sociopsychological, and gendered vulnerability drivers that shape decisions on whether to adopt, dis-adopt, or not adopt CSA technologies. This emanates from a focus on CSA as solving the dilemma of climate change through technical fixes to increase food production. The provision of technological solutions for resilience requires consideration of their social implications, the absence of which has resulted in growing concern over the observed adoption paradox. Failure to address the underlying gender inequalities and vulnerabilities may have ramifications for resilience-building for smallholder farmers. Additionally, an existing understanding of CSA adoption is framed within a simplistic linear approach, which is insufficient when gender and resilience-building dimensions are brought into consideration. Thus, this study was conducted with the aim to explore the application of a resilience perspective to CSA to address the underlying gender inequality and gendered vulnerability in order to improve CSA adoption by diverse men and women smallholder farmers.

The shortcomings of CSA have also been linked to the issue of a conceptual misnomer, arising from a general conceptualization of CSA that includes policies, technologies, and practices at farmer, landscape, and ecosystem levels (Lipper et al. 2014). While some literature labels CSA as an already compromised concept pushing a hegemonic agenda for the developed countries (Taylor 2018), the concept has the potential to address some of the challenges African societies face with climate change. Having considered the concerns, this study situates CSA adoption assessment at a farmer level and with a deliberate focus on technologies and practices that farmers have to adopt. Some scholars highlight the need for alternative frameworks to address the shortcomings of CSA (Taylor 2018; Glover et al. 2019). This gives currency to the application of resilience thinking in CSA adoption, and a reconnaissance of CSA that leverages on its relationship with disaster risk reduction (DRR) can help address some of the highlighted shortcomings.

The Concept of CSA in a DRR Context

The second pillar of CSA is "resilience-building and adaptation," which underpins the interconnectedness of CSA and DRR (FAO 2013). This relationship paves the way for the application of a DRR lens to CSA in order to explore opportunities for improving CSA adoption by smallholder farmers. The climate-related risks and disasters affecting smallholder farming as already outlined in the introduction of this chapter give credence to such an approach. Moreover, at a farmer level, the demarcations between adaptation, resilience-building, and DRR are indistinct, as farmers are more concerned with surviving each disaster event.

A DRR approach to CSA draws attention to issues of vulnerability reduction, while CSA implementation in smallholder farming provides a vehicle to deliver both risk reduction and adaptation simultaneously (FAO 2013). A DRR approach to CSA could help resolve some of the shortcomings of CSA as highlighted in the preceding section. Greater strides have been made in DRR than in CSA, for example, in terms of the appreciation of resilience-building, indigenous knowledge systems, and the application of socio-ecological system concept to understand resilience-building and community-based participation (FAO 2013; Coetzee et al. 2016). Therefore, CSA could draw from progress made in DRR this far as a way of resolving the adoption challenges. Unfortunately, there has been minimal scholarly interrogation of CSA from a DRR perspective. Yet, at a time when increased climate risk threatens to wipe out the development gains made in agriculture so far, such a consolidated approach could better cross-examine CSA adoption. Furthermore, the relationship of DRR and resilience provides basis to interrogate CSA adoption from a disaster resilience perspective.

Disaster resilience is framed as an *ability*, where a system and its units are able to anticipate, absorb, accommodate, and recover from a disturbance by bouncing back or bouncing forward timeously and efficiently (Manyena et al. 2011). The system and its units may have the ability to change without loss of basic structure and functions, or self-organize, attaining incremental capacity to learn, adapt, and change through absorptive, adaptive, or transformative capacities (Béné et al. 2016). When smallholder farmers make decisions to adopt CSA technologies and practices, that is essentially indicative of their aspirations to be resilient to climate vagaries. Resilience of a system or its units, which in this study were individual farming households in a farming system, is better appreciated by considering resilience principles. This includes maintenance of redundancy and diversity, management of intra-system connectivity, feedbacks, promotion of social learning, participation and inclusion, embracing polycentricity, and understanding that agricultural systems are complex adaptive systems (Carpenter et al. 2012; Coetzee et al. 2016). Therefore, in assessing CSA adoption challenges from a resilience perspective, we argue that these resilience principles can be applied to assess barriers to CSA adoption and may help identify required improvements to build the resilience of farming households and communities.

Absorptive resilience is when households are able to contend with the negative effects of climate disasters through persistent coping and resistance, without any distinct changes to function or structure (Bennett et al. 2014), for example, when households cope with a drought through humanitarian interventions such as food aid distribution. Adaptive resilience is when the agricultural system or its units have the ability to learn from acquired or experiential knowledge and to make adjustments in response to disasters (Walker et al. 2004). In adaptive resilience, the aim is to make adjustments for continued functioning within a household or system. Transformative resilience refers to the capacity for change in structure and function of the system or households owing to disturbance. Transformation is more concerned with changes made in behaviors, cultural ethos, stereotypes, institutions, and policy direction (Walker et al. 2004). Thus, transformation depends on interrogation of the status quo and advocating for pragmatic changes in structure or function to be instituted. Adaptation and transformation are long-term and essential dimensions of resilience from a development standpoint. It is important to bear in mind that the three dimensions should not be pursued separately in linear fashion. It is important to harness the existing synergies among them (Béné et al. 2016).

Accordingly, for the majority at-risk rural smallholder farmers, CSA offers a pragmatic relevant conduit to pursue resilience. The assorted CSA options (see Table 1) contribute or have the potential to contribute to the three resilience dimensions, so it is worth mentioning that CSA implementation and policies should not elevate any one dimension and subordinate the others. Rather, in building on the synergistic relationships of absorption, adaptation, and transformation, CSA can help smallholder farmers and their systems become resilient.

In the context of smallholder farming in developing regions where CSA is promoted, it is key to recognize the heterogeneity of the farmers (Khoza et al. 2019).

The diversity among smallholder farmers draws attention to existing inequalities that relate to vulnerability and shape power, agency, ownership and control of resources, decision-making, and participation within farming systems (Ensor et al. 2018). This magnifies the need for resilience-building in CSA to pay attention to the skewed landscape within which CSA adoption decisions have to be made by different farmers. Ultimately, this mandates that over and above absorptive and adaptive resilience, transformation is required in CSA. This compels the interrogation of existing social imbalances that determine whether a smallholder farmer will adopt, dis-adopt, or not adopt CSA.

Methodology

An exploratory sequential mixed method design (Teddlie and Tashakkori 2009) was applied in Chikwawa, Malawi, and Gwembe, Zambia, to gather empirical data at a local level where smallholder farmers interface with climate-related disasters and

Table 1 Climate-smart agriculture options

CSA options	Examples
Crop management	Intercropping
	Crop rotation
	Crop diversification
	Improved seed varieties
	Value chains and marketing
	Improved postharvest storage
	Agro-processing
Livestock management	Fodder crops
	Feedlots
	Improved breed
	Rotational grazing
	Grassland restoration and conservation
Soil and water management	Basin/mechanized conservation farming
	Solar-powered irrigation
	Rehabilitation of degraded landscapes
Agroforestry	Woodlots
	Fruit trees
	Nitrogen-fixing trees
	Multipurpose trees
Integrated food-energy systems	Biogas stoves
	Energy-saving stoves
Infrastructure	Roads
	Housing
	Mobile network
Access to climate information	ICT platforms/information hubs
Fisheries	Aquaculture
	Capture fisheries

Adapted from FAO (2013)

where resilience-building is essential. The study was conducted between 2017 and 2019, with data collection in February and March 2018. The initial phase entailed qualitative data collection through semi-structured face-to-face interviews from purposively selected key informants at district level and focus group discussions (FGDs) at ward level. A total of 16 interviews and 6 FGDs were conducted (3 in each country: men only, women only, and mixed men and women). Thematic qualitative data analysis informed the design of an instrument used in quantitative cross-sectional data collection. In the quantitative cross-sectional survey, a total of 102 smallholder farmers were interviewed, 51 from each study site. The cross-sectional survey served to explore the generalizability of the themes established from the qualitative findings. In order to capture the perspectives and contexts of the gender dimensions in CSA adoption, the study placed emphasis on qualitative findings. This is in line with the methodological provisions of a mixed method research design (Teddlie and Tashakkori 2009). The quantitative data was analyzed with SPSS version 26 for descriptive statistics that established the distribution and trends.

Findings

Ownership of Land

The findings revealed that in Chikwawa, the average land owned by men household heads was 1.4 ha, while for women, it was 0.7 ha. In Gwembe, 40% of women-headed households who indicated they rented land did so for CSA purposes. While men household heads generally rented land in addition to what they owned, the women rented land because they were landless. Land ownership by women in Africa is a contentious issue (Doss et al. 2015), and these findings call for renewed effort to address the issue. Land ownership influences the adoption of agricultural technologies and practices; therefore, if CSA is to contribute to resilience-building, there is a need for equal distribution of land as a starting point toward equitable resilience (Matin et al. 2018).

Participation in CSA Technology Development

The qualitative findings established that CSA technology development occurred in a top-down manner, with smallholder farmers not engaged in technology development as they are generally considered as recipients "who receive your technology you have developed for them" (NGO respondent, Chikwawa). In both study sites, field days and demonstration plots were identified as opportunities for farmer participation in technology development. However, respondents acknowledged that even these events were top-down as they mainly showcased technologies that had been developed *for* the farmer, and technologies developed *with* the farmers' involvement were rare, if any. Currently, no mixed approach to CSA technology development that comprises technologies developed *for* and *with* the farmers is being considered. This may be due to the perception that farmers are technology recipients, as reflected by some interviewees: "they cannot contribute anything in technology development... what do farmers know that they can contribute in CSA?" (Government Department Respondent, Gwembe).

These sentiments were corroborated by quantitative findings that established that there was minimal participation among farmers in technology development irrespective of gender. In Chikwawa, 25% of the households, with over 70% of these being male household heads, stated that they had been involved in meetings when conservation farming and irrigation schemes were first brought to their communities. In Gwembe, 11% of the farmers acknowledged participation in similar meetings. At both study sites, those who participated in meetings went on to become adopters as they benefited from the respective CSA projects.

However, this is insufficient as participation should also include problem identification, evaluation of options, and eventual selection of technologies that farmers know will address their problems. This would leave room for the consideration of indigenous knowledge systems, which can also be considered as alternatives to solve the problems farmers face. For example, in Chikwawa, farmers shared how they used a fish broth to control the fall armyworm before pesticides were available. Scientific research could be incorporated to explore how indigenous knowledge can be harnessed in CSA.

CSA Options Available for Farmers

The qualitative findings established that conservation farming was the major form of CSA that farmers have adopted, and confirmation from quantitative findings revealed that 100% of farmers practicing CSA at both sites stated that they used improved seed varieties (ISVs) and soil moisture conservation techniques. At both study sites, less than 40% of sampled households were engaged in more than one form of CSA. In Chikwawa, other forms of CSA included small-scale irrigation schemes, while in Gwembe, a new aquaculture project was at inception stage at the time of data collection. In Gwembe, less than 20% of interviewed farmers had also been previously involved in improved livestock breed projects. However, the qualitative findings showed that there were concerns that conservation farming alone was insufficient in addressing farmers' needs as explained by practitioners:

> . . .we know that crop production is always vulnerable, we also need to bring in livestock for the farmers, to help them when crops fail. . .especially goats which they can sell when crops fail. (Respondent from Government Department, Chikwawa)

The quantitative findings showed that livestock ownership differed between male and female heads of households. At both sites, married men owned the most cattle, with average cattle ownership in Chikwawa being two heads, while in Gwembe, it was eight. More female household heads owned cattle in Gwembe than in Chikwawa, 16% and 7%, respectively. When these trends are viewed from an intersectionality perspective, intersection of gender with education and wealth status can be noted as the women who owned cattle in Gwembe were predominantly retired professionals who were categorized as better-off in the community wealth rankings.

CSA Goals for Farmers

In Chikwawa, qualitative findings established that the intended CSA outcomes of improved agricultural productivity and resilience-building were not being achieved by means of the CSA options available to farmers. Evidence of these shortcomings of CSA was linked to humanitarian food assistance. Qualitative findings indicated that there was no major difference in terms of food security between CSA farmers and those who were not involved in any form of CSA because "we see it when it comes to food aid, they all need assistance because they will be all food insecure" (Respondent from Government Department, Chikwawa). More concerning were sentiments from non-adopters who indicated a lack of motivation to adopt available CSA options because "we are all the same, CSA does not make them any better than us" (Discussants in mixed gender FGD, Chikwawa). These findings were confirmed by the quantitative survey where 100% of farmers who adopted conservation farming also reiterated that they benefited from food aid each year because of low crop yields. Farmers using ISVs raised concerns about their susceptibility to the fall armyworm, claiming that their traditional open-pollinated varieties (OPVs) were better resistant.

In Gwembe, a different scenario emerged when assessing whether CSA options were able to contribute to food security and resilience-building. Qualitative findings established that while yields increased through conservation farming, there were postharvest crop losses as farmers could not sell their surplus anywhere. Quantitative findings confirmed these sentiments as 100% of the farmers who were practicing conservation farming techniques were utilizing less than half of their arable land for CSA to "avoid high yields that they would still lose through spoilage," as farmers concurred during household survey in Gwembe.

Discussion

Gender-Equitable Resilience in CSA Adoption

The findings presented in this chapter show that men and women farmers may not be realizing benefits of the CSA activities they are involved in. Findings further show that currently, CSA is not contributing toward the resilience of farmers as they are still prone to food insecurity, often relying on food aid to see them through to the next season. This creates a dependency syndrome and demotivation farmers in CSA adoption. Moreover, dominance of conservation farming leaves farmers vulnerable to climate hazards that have a negative effect on crop production. Findings illuminate the insufficiencies of current CSA and gaps that continue to hinder CSA adoption, especially among women farmers. This chapter accentuates that a resilience framing of CSA leaves room for broader consideration of the decision-making context within which smallholder farmers live. On that basis, a normative gender-sensitive CSA adoption model (Fig. 1) is proposed.

The aim of the framework is to provide a normative approach to improve CSA adoption, especially by diverse women smallholder farmers in developing regions, considering their central role in farming activities. The framework is conceptualized from a resilience viewpoint, enabling a more holistic approach to the issues that may enhance decision-making by different groups of farmers, especially diverse women smallholder farmers. There is a need for gender transformation at various CSA implementation levels, starting at household level up to national and global levels. Transformation requires various strategies and enablers to be put in place to create

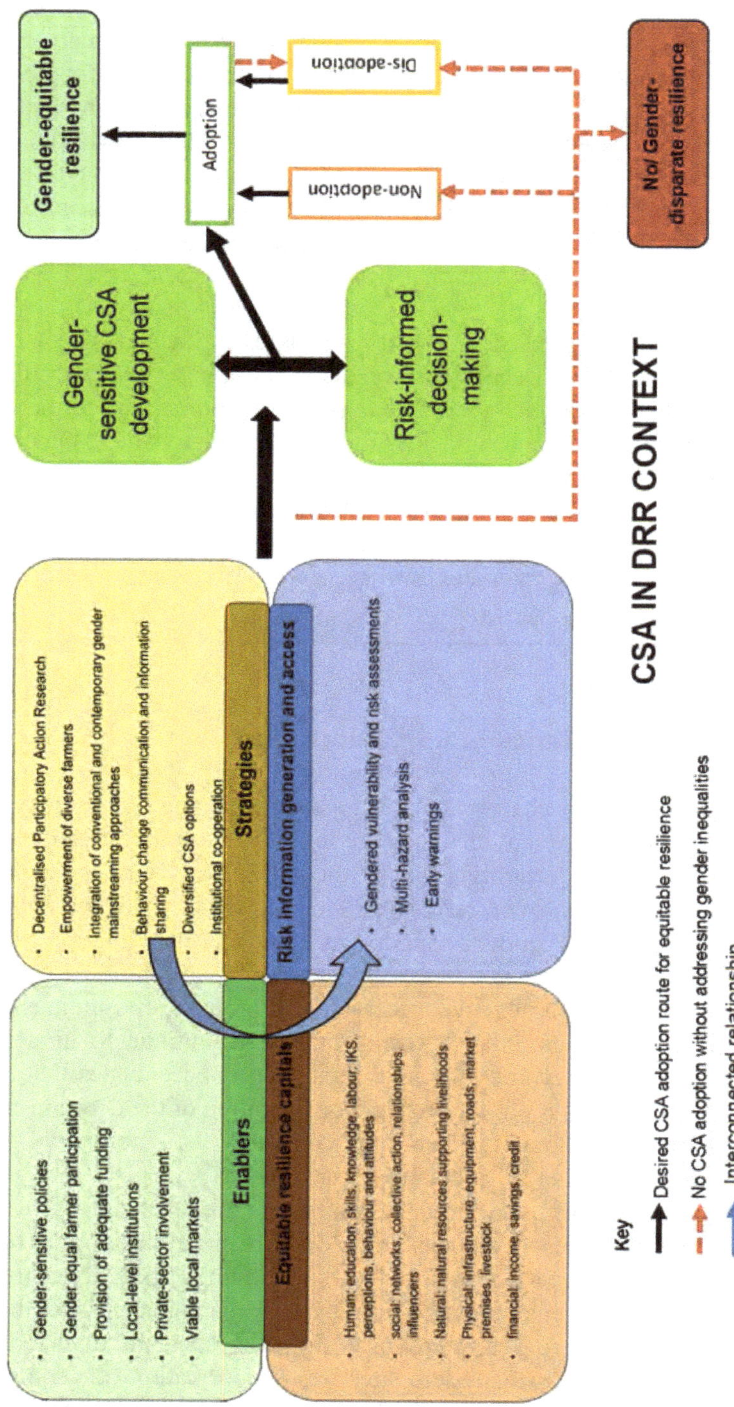

Fig. 1 Proposed gender-sensitive CSA adoption framework

equality and address gendered vulnerability, which should potentially result in improved CSA adoption at household level. The gender-sensitive CSA adoption framework comprises various interconnected components that should be used from a gender perspective throughout, aimed at transformation toward more egalitarian-resilient societies. In the proposed framework, the desired adoption route likely to help achieve gender-equitable resilience is illustrated with black arrows, and the undesired route likely to result if gender disparities are not addressed is shown with red dotted arrows. The blue curved arrow shows that the enablers, strategies, gender equality, and risk information are all interconnected, interacting to inform gender-sensitive technology development and risk-informed decision-making. The framework components are discussed in the following subsections.

Enablers for CSA Adoption

Gender-Sensitive Policies

The findings of this study show that there are preexisting gender inequalities at household level which may be perpetuated by gender-blind CSA implementation. This chapter suggests that an improvement in CSA adoption by women farmers would require gender-sensitive policies to ensure that issues of gender inequality are addressed to achieve gender parity. This requires a holistic assessment of CSA that will consider implementation strategies and resilience capitals, not just limited to the technological benefits of CSA. There is need for policies directly or indirectly linked to CSA to be assessed for their implications on different genders. This should include land tenure systems, marriage and property inheritance laws, technology development, as well as economic empowerment, which all affect CSA adoption decisions (Doss et al. 2015; Khoza et al. 2019).

Gender-Equal Farmer Participation

CSA presents various opportunities where farmers should be engaged for active participation in CSA technology development. However, study findings showed that currently, farmers' participation in CSA is mainly as recipients of already developed technologies and CSA information. CSA is characterized by top-down approaches which may fail to pay attention to critical gender issues that hinder adoption. This study reiterates the need for CSA implementation to ensure equal participation of farmers in technology development and identification of CSA options to adequately meet the resilience needs of diverse farmer categories. There should also be gender-equal participation in the co-creation of knowledge through research in gendered risk assessments, vulnerability assessments, and multi-hazard analysis. Gender-equal participation of farmers will likely assist with the identification of gender-differentiated barriers of CSA adoption and opportunities that can be harnessed to improve adoption across different genders.

When smallholder farmers are given equal opportunities to participate in various aspects of CSA, this is likely to also bring to the fore critical contextual gender issues and to facilitate transformation. Gender-equal participation may potentially enable bottom-up engagement in CSA, where farmers can also contribute their knowledge and experiences. This is especially important when considering the role of indigenous knowledge systems in CSA. Ultimately, equal participation of farmers allows CSA to engage with their various realities, ensuring farmers have a voice in the design and ownership of CSA projects and technologies. When farmers are given the space to participate in various components of CSA, not just as recipients, they are more likely to adopt CSA. This will also enhance sustainability of CSA in communities.

Provision of Adequate Funding

There is no doubt that technological requirements of CSA are likely to be costly and beyond the reach of many individual farming households. For instance, capital investment for some CSA options, such as irrigation schemes and aquaculture, may be costly. This means that at a higher national and global scale, there is a need to improve funding for CSA projects. This can be achieved through multiple funding streams. For example, at a national level, countries should fulfil commitments of the Malabo Declaration that states that African governments should allocate 10% of their public spending to agriculture (AU 2014). Other funding sources could be explored through other government sectors. For instance, on the basis of its relationship with DRR, disaster risk management (DRM) departments could also provide part of the funding required for CSA. Similarly, at a global level, multiple sources may be explored apart from specific CSA projects, such as resilience-building, climate change adaptation, or DRR funds.

Importantly, CSA funding should ensure that funds provided address the resource needs of local-level institutions, such as the provision of vehicles, information and communication technology equipment, and recruitment of more extension agents. This will likely improve quality of gender-specific contact extension services provided to the farmers. This may also compel strategic direction toward the integration of local-level institutions.

Local-Level Institutions

Strong operational relationships among local government and nongovernmental institutions are required to facilitate gender-sensitive CSA adoption. According to Carpenter et al. (2012), polycentricity is key to resilience, often helping to promote connectivity within systems and facilitating learning. This is key in CSA where diverse institutions need to work together. These institutions are the first-level responders to different hazards affecting smallholder farmers; hence, their cooperation is necessary to improve adoption of diverse CSA options. Local-level institutional cooperation and collective action remain essential in the provision of extension services, information dissemination, and facilitation of gender transformation in communities. Ultimately, a polycentric approach enhances the delivery of CSA with precision and efficiency to meet gender-specific farmer resilience needs.

Private Sector and Viable Markets

The findings reveal that one major drawback of CSA adoption by both men and women farmers is the lack of improved quality of life as a result. This was connected to a lack of economic opportunities, resulting in farmers failing to earn meaningful income from the sale of surplus produce. This chapter recommends that innovative strategies be implemented to involve the private sector in CSA to ensure win-win scenarios for farmers and business. CSA adoption is negatively affected by unviable local markets, and an enabling environment for CSA adoption should consider creation of viable local markets where farmers can buy and sell CSA inputs and outputs. This may help create a thriving local economy and increase the income earned from CSA to meet household needs.

Strategies to Improve CSA Adoption

Decentralized Participatory Action Research

The gender disparities identified in the study magnify the need for CSA adoption to be informed by participatory action research (PAR), which can be achieved with an enabling environment for gender-equal participation of smallholder farmers. PAR may facilitate engagement with farmers, giving them a platform to share their experiences in gender issues that demotivate them from adopting CSA or drive them to discontinue CSA. Furthermore, PAR should be decentralized, allowing research to be conducted at the local epicenter of climate disasters. The strength of PAR in driving CSA adoption is the recognition of farmers as both sources and consumers of knowledge, with their involvement in research tapping into local knowledge, perspectives, and realities. At the same time, they are able to use the information from PAR to inform their CSA adoption decisions. Decentralization of PAR to local level should be gender-sensitive, identifying best ways to examine challenges and opportunities for specific groups of farmers in CSA. Farmer participation will illuminate sociopsychological behaviors, attitudes, and perceptions of diverse groups, ensuring that precise and complete information equips farmers' decision-making. PAR is also essential in creating a platform for behavior change communication and information sharing.

Diversity of Livelihoods and CSA Options

Considering CSA from a resilience perspective magnifies the need for CSA to move beyond dominance of conservation farming as revealed in this study. If CSA is to contribute to resilience of diverse smallholder farmers, then there is a need to provide diversified CSA options in addition to conservation farming. Diversified CSA options ensure redundancy so that in the face of a climate-related disaster affecting one component of the farming systems, farmers have other alternatives to rely on (Carpenter et al. 2012). The dominant focus on conservation farming could help explain protracted food insecurity and vulnerability, with farmers often relying on food aid assistance. Therefore, a resilience lens in CSA advocates for transformation toward consideration of other livelihoods and CSA options. This may include income generation through sale of improved livestock breeds, honey from apiculture, or fish from aquaculture, among others. Diversity and redundancy should improve the resilience of farmers, and gender should be considered to assess which CSA options would be relevant to each category of farmers.

Empowerment of Diverse Women Farmers

In view of the study findings, this chapter accentuates the need for the empowerment of communities in general and women in particular to equip them to be able to articulate their resilience needs and to demand more space to participate in different aspects of CSA. The heterogeneity of women smallholder farmers is suggestive of their corresponding diverse resilience needs. This is essential when considering issues of economic empowerment in CSA. CSA adopters have not been able to derive tangible economic benefits, yet this is one of the goals of CSA. Empowerment of farmers means they will participate in technology development, contributing toward defining relevant CSA options they need, and they will participate in decision-making at various levels from intra-household level going up.

Empowerment in CSA should ensure that women farmers can share their experiences, practice autonomy and agency, and be able to collectively come together to tackle structural bottlenecks affecting their adoption decisions. However, this requires a transformation from traditional gender mainstreaming approaches that have shaped empowerment efforts in the past, toward an integrated approach that also considers contemporary approaches such as intersectionality, African feminisms, and positive masculinity (Arndt 2002; Davis 2008). Studies have shown insufficiencies of traditional gender mainstreaming approaches in addressing gender inequality and patriarchy in agriculture (Khoza et al. 2019). The integration of traditional and contemporary approaches may compensate for the weaknesses of each approach applied on its own. Empowerment should address practical gender needs while also ensuring that attention is paid to structural gender issues that may hinder especially women household heads from adoption of CSA. Empowerment should also pave way for participation and inclusion of farmers, especially women, in the various aspects of CSA as explained in earlier sections. This remains an essential vehicle of transformation.

Gender-Equitable Resilience Capitals

Based on the study findings, this chapter highlights that a resilience framing of CSA adoption compels consideration of gender inequality and gendered vulnerability with regard to access to and control and ownership of resilience capitals (Mayunga 2007). The gender constructions that determine who owns, has access, and controls resilience capitals should be assessed in CSA as they shape farmers' adoption decisions. In order to achieve resilience-building through CSA, there is a need for deliberate strategies aimed at establishing gender equality and equity in the ownership, control of, and access to social, natural, physical, financial, and human resilience capitals. This will require CSA to engage with disparities and to improve especially the ownership of resilience capitals such as farming equipment, livestock, land, and finance by women farmers. This will not only enable them to cope with climatic disturbances but will also ensure they are equipped to build back better or bounce forward from each disturbance. Paying attention to resilience capitals also helps illuminate key vulnerability issues that dispose farmers to either dis-adoption or non-adoption of CSA. Creating gender equality and equity in resilience capital ownership will require innovation in tackling the socioculturally entrenched patriarchal systems and women's subordination, and contemporary gender mainstreaming approaches may be useful in this regard.

While addressing identified gender inequality issues may not be the primary mandate of agricultural departments, a resilience framing emphasizes polycentricity and collective action. Other development actors need to be involved in CSA, such as gender, DRM and community development departments, NGOs, women's rights activists, and local leaders. These structures already exist at a local level, although agriculture departments may need to lead the integration to ensure that the expertise of various groups is channeled toward addressing inequality and vulnerability in pursuit of resilience.

Risk Information: Generation and Access

Any attempt to improve CSA adoption requires strategies to ensure supply of adequate information to aid farmers during decision-making. Collective action, participation, and inclusion are key to the generation of risk information. Processes to generate risk information are undertaken by governments, NGOs, and donor agencies in many countries. These are usually in the form of vulnerability and risk assessments, as well as hazard analysis. However, there is a need to move beyond simple gender-disaggregated data generated in these processes to critically engage with the gender implications of collected data in terms of resilience-building. Risk information is not only useful to technocrats and practitioners, but farmers should also have access to the information for decision-making. Knowledge is required to make informed decisions. Across different gender groups, its creation and acquisition is important in decision-making. The proposed framework advocates for the involvement of farmers in knowledge co-creation, which will harness valuable indigenous knowledge, especially with regard to climate hazards and early warnings. This means attention should be paid to access to gender-sensitive risk communication. Gender-sensitive risk information is also requisite in development of gender-sensitive CSA technologies.

A systemic approach helps appreciate that CSA adoption decisions are not only based on technological benefits of CSA options. Farmers consider other risks that affect their resilience capitals negatively or positively within the wider system context. For instance, for many communities, disease epidemics such as HIV/AIDS remain a health risk that threatens agricultural labor provision in households. Therefore, any adoption improvement strategy should engage farmers to identify what other risks they face in their contexts, and this may be achieved through gender vulnerability and risk assessments, as well as multi-hazard analyses that should endeavor to obtain in-depth qualitative perspectives on systemic risks.

Risk-Informed Decision-Making

Adoption decisions of men and women smallholder farmers are influenced by various factors depending on their gender roles (Khoza et al. 2019). Importantly, decision-making for men and women household heads should be viewed within the multifaceted context in which decisions are made and have to be risk-informed. There is a need to acknowledge different factors and drivers that shape decision-making for different genders. A resilience framing of CSA accommodates risk-informed decision-making (RIDM) even at smallholder farmer level (Weichselgartner and Pigeon 2015). RIDM acknowledges that decision-making is not in simple linear fashion as traditionally understood. It is a more comprehensive analytical approach that interrogates and seeks to understand complex interactions between people, risks, hazards, and systems. Risk-informed decisions pay attention to qualitative information from gender-differentiated risk assessments (Gardoni et al. 2016), narratives, and realities that shape decisions by different farmers. However, Apostolakis (2004) caution against the exclusive use of risk assessments to inform decisions, necessitating a more consolidated approach where gender vulnerability assessments and multi-hazard analyses can also feed into decision-making.

Gender-Sensitive CSA Technology Development

Findings of this study showed that smallholder farmers, irrespective of gender, were not directly involved in the development of CSA technologies. Rather, technology development was a top-down process where farmers were viewed as recipients. However, this chapter argues that if CSA adoption is to be improved, there is a need for farmers to participate in technology development. CSA technology development should be two-way, with provision for consideration and development of local farmer innovations for further scaling up. Development and dissemination of CSA technology should be participatory to generate and manage perspectives that may determine adoption decisions made especially by the women farmers. CSA technology development should therefore be informed by the gender analyses that recognize gender roles and interactions with technology in relation to culture, behaviors, attitudes, and social influences (Ngigi et al. 2018; Khoza et al. 2019). Development of CSA technology should appreciate and address any underlying disparate distribution of resilience capitals. Failure to consider these underlying factors and corresponding strategies to address them may manifest as low adoption of CSA by women farmers.

Additionally, through gender analyses, CSA technology development can consider existing and projected changes in gender roles. CSA technology may seek to improve current gender roles, or transform them, depending on identified inequalities and farmer needs (Nyasimi and Huyer 2017), where technologies can be developed to help bridge the gender productivity gap and contribute to equitable resilience across the heterogeneity of smallholder farmers. For instance, women in men-headed households and women who were household heads lamented labor demands of basin conservation farming which they felt increased their workload, while they have other reproductive and community roles too. Moreover, caution should be exercised to ensure that CSA does not reinforce gender stereotypes, for instance, when CSA projects target women only for energy-saving stove distribution.

Critical to gender-sensitive technology development is the cost of CSA technologies. Women who are already less economically empowered than men are less likely to be able to afford costly new CSA technologies. CSA focus should also be on women's economic empowerment. Ultimately, rural women need appropriate CSA technologies that can transform their contexts and realities where necessary, helping them to become more resilient. This can be achieved by engaging the diverse groups of women to establish their practical and structural gender needs. Gender-sensitive CSA technology development should be as pragmatic and transformative as possible in pursuit of resilience.

Operationalization of the Framework

This chapter suggests that utilitarian value of the framework lies in its ability to identify and confront issues of inequality and social disparities in a broader context, which may pave way for decision-making that favors CSA adoption by smallholder farmers. Operationalization of this framework should start at a district level and bring together communities and experts from diverse disciplines such as agriculture, DRM, climate change, gender, community development, local leaders, businesses, weather services, research institutions, and NGOs. Most of these disciplines are already represented at district level, although there is a need to transition toward collective integrated operations. The agriculture department may maintain the leadership and coordinating mandate, ensuring representation and multidirectional participatory engagement, communication, and information dissemination. The use of the framework can then feed into large-scale administrative processes at the provincial and national levels. Some components of the framework are already addressed by ongoing activities, such as vulnerability assessments and hazard and risk assessments. However, a gender lens should be applied in these processes, which should include smallholder farmers in their diversity, and findings from assessments should be used to inform all DRR components, not just for response through humanitarian food assistance.

The proposed framework is worth exploring as it derives value from the participatory nature of its formulation and has a strong focus on social dimensions in CSA adoption. As such, it addresses some of the gaps in current appreciation of CSA adoption, which seems to elevate the technological merits of CSA at the expense of the equally important social dimensions. This framework's ingenuity also lies in that it speaks to the insufficiencies of a linear approach to CSA. Challenges may arise in that the framework was developed independent of any existing CSA project, which means that its uptake by different institutions is not guaranteed. Nevertheless, it does present a normative approach to improving CSA adoption so that men and women smallholder farmers can be enabled to "build back better equally, leaving no-one behind," which should form the core of resilience and sustainable livelihood outcomes in Africa.

Key Lessons, Study Limitations, and Future Research Prospects for CSA

Several key lessons can be drawn from this study that applied a resilience framing to the interrogation of gender dimensions in CSA adoption. Firstly, at local level where loss and damage from each disaster event worsen food insecurity and poverty, there is an amplified urgency for gender-equitable resilience-building through CSA. This compels improvement of CSA adoption by at-risk smallholder farmers. Drawing on resilience principles may give strategy direction in CSA implementation at the local level. Resilience principles such as the maintenance of redundancy and diversity, management of intra-system connectivity, feedbacks, promotion of social learning, participation and inclusion, embracing polycentricity, and understanding that agricultural systems are complex adaptive systems all play an essential role in pursuit of gender-equitable resilience in CSA.

Secondly, there remains need to address gender inequality as it stands in the way of CSA adoption and inhibits the successful pursuit of resilience. Areas of focus should include the promotion of active and equal participation of smallholder farmers, especially vulnerable, at-risk women in CSA technology development. CSA technologies should not merely be developed *for* the smallholder farmers, but *with* them, creating space for consideration of their innovations and indigenous knowledge systems to suit their gender-specific resilience needs.

Lastly, it is essential to shift from risk-based decision-making toward risk-informed decision-making, starting at household level right up to institutional level where CSA support is provided. This is dependent on an appreciation of the fact that CSA adoption decision-making is not linear and only based on the technological benefits of CSA in the face of climate change, but rather occurs within a multifaceted decision-making context, which may differ along gender lines.

Further studies on CSA can adapt and replicate the study methodology in similar developmental contexts and disaster profiles. This study did not test the applicability of the proposed gender-sensitive CSA adoption framework. Future studies can test the applicability of the proposed framework in any of the Southern African countries or in other developing regions. The key is sensitivity to site-specific contexts on issues of gender, inequality, power, and agency. CSA adoption itself is a difficult variable to assess by means of a cross-sectional survey. Longitudinal studies can be considered in future research.

Conclusions

The assessment of the gender dimensions of CSA adoption by smallholder farmers anchored in resilience thinking enriches understanding of the inherent challenges specifically faced by women farmers. The study highlights several factors that drive gender-differentiated vulnerability and contribute to low CSA adoption by at-risk women farmers. These factors include gender disparities in ownership of resilience capitals, limited participation of farmers in CSA technology development, narrow CSA options that can be adopted by farmers, and the limited tangible benefits realized from CSA. There is a need to improve the adoption of CSA by diverse smallholder farmers and to meet their gender-specific resilience requirements in the face of climate-related disasters. This can be achieved through the creation of an enabling environment and implementation of strategies that can facilitate gender-sensitive CSA technology development and the equal participation of smallholder farmers. The farmers should be assisted to make risk-informed decisions by improving generation and access to risk information through participatory action research. Gender equity and equality in the ownership of resilience capitals can contribute to gender-equitable resilience in smallholder farming. In conclusion, this study reiterates the importance of a gendered approach to CSA to improve adoption by smallholder farmers who directly interface with inclement climate hazards.

References

African Union (AU) (2014) Implementation strategy and roadmap to achieve the 2025 vision on CAADP. African Union, Addis Ababa

Apostolakis GE (2004) How useful is quantitative risk assessment? Risk Anal 24(3):515–520

Arndt S (2002) Perspectives on African feminism: defining and classifying African-feminist literatures. Agenda 17(54):31–44

Arslan A, Asfaw S, Cavatassi R, Lipper L, McCarthy N, Kokwe M, Phiri G (2018) Diversification as part of a CSA strategy: the cases of Zambia and Malawi. In: Lipper L, McCarthy N, Zilberman D, Asfaw S Branca G (eds) Climate smart agriculture: building resilience to climate change. Springer, Cham

Asfaw S, Bishop-Sambrook C, Diei Y, Firmian I, Henninger N, Heumesser C, Huyer S, Kristjanson P, Lefter C, Lehel S (2015) Gender in climate-smart agriculture: module 18 for gender in agriculture sourcebook. World Bank Group, Washington, DC

Béné C, Headey D, Haddad L, von Grebmer K (2016) Is resilience a useful concept in the context of food security and nutrition programmes? Some conceptual and practical considerations. Food Sec 8(1):123–138

Bennett E, Carpenter S, Gordon L, Ramankutty N, Balvanera P, Campbell B, Cramer W, Foley J, Folke C, Karlberg L (2014) Toward a more resilient agriculture. Solutions 5(5):65–75

Carpenter S, Arrow K, Barrett S, Biggs R, Brock W, Crépin A-S, Engström G, Folke C, Hughes T, Kautsky N (2012) General resilience to cope with extreme events. Sustainability 4(12):3248–3259

Chandra A, McNamara KE, Dargusch P (2017a) Climate-smart agriculture: perspectives and framings. Clim Pol 18(4):1–16

Chandra A, McNamara KE, Dargusch P (2017b) The relevance of political ecology perspectives for smallholder climate-smart agriculture: a review. J Polit Ecol 24(1):821–842

Coetzee C, Van Niekerk D, Raju E (2016) Disaster resilience and complex adaptive systems theory: finding common grounds for risk reduction. Disaster Prev Manag 25(2):196–211

Collins A (2017) Saying all the right things? Gendered discourse in climate-smart agriculture. J Peasant Stud 45(1):1–17

Davis K (2008) Intersectionality as buzzword: a sociology of science perspective on what makes a feminist theory successful. Fem Theory 9(1):67–85

Diao X, Hazell P, Thurlow J (2010) The role of agriculture in African development. World Dev 38 (10):1375–1383

Doss C, Kovarik C, Peterman A, Quisumbing A, van den Bold M (2015) Gender inequalities in ownership and control of land in Africa: myth and reality. Agric Econ 46(3):403–434

Ensor J, Forrester J, Matin N (2018) Bringing rights into resilience: revealing complexities of climate risks and social conflict. Disasters 42(2):287–305

Food and Agriculture Organisation (FAO) (2013) Climate-smart agriculture sourcebook. FAO, Rome

Gardoni P, Murphy C, Rowell A (2016) Risk analysis of natural hazards: interdisciplinary challenges and integrated solutions: risk analysis of natural hazards. Springer, Cham

Glover D, Sumberg J, Ton G, Andersson J, Badstue L (2019) Rethinking technological change in smallholder agriculture. Outlook Agric 48(3):169–180

Khoza S, Van Niekerk D, Nemakonde LD (2019) Understanding gender dimensions of climate-smart agriculture adoption in disaster-prone smallholder farming communities in Malawi and Zambia. Disaster Prev Manag 28(5):530–547

Kpadonou RAB, Owiyo T, Barbier B, Denton F, Rutabingwa F, Kiema A (2017) Advancing climate-smart-agriculture in developing drylands: joint analysis of the adoption of multiple on-farm soil and water conservation technologies in West African Sahel. Land Use Policy 61:196–207

Lipper L, Thornton P, Campbell BM, Baedeker T, Braimoh A, Bwalya M, Caron P, Cattaneo A, Garrity D, Henry K (2014) Climate-smart agriculture for food security. Nat Clim Chang 4 (12):1068–1072

Lipper L, McCarthy N, Zilberman D, Asfaw S, Branca G (2018) Climate smart agriculture: building resilience to climate change. Springer, Cham

Manyena B, O'Brien G, O'Keefe P, Rose J (2011) Disaster resilience: a bounce back or bounce forward ability? Local Environ 16(5):417–424

Matin N, Forrester J, Ensor J (2018) What is equitable resilience? World Dev 109(9):197–205

Mayunga JS (2007) Understanding and applying the concept of community disaster resilience: a capital-based approach. In: Summer academy for social vulnerability and resilience building, Munich

Müller C, Waha K, Bondeau A, Heinke J (2014) Hotspots of climate change impacts in sub-Saharan Africa and implications for adaptation and development. Glob Chang Biol 20 (8):2505–2517

Murray U, Gebremedhin Z, Brychkova G, Spillane C (2016) Smallholder-farmers and climate smart agriculture: technology and labor-productivity constraints amongst women smallholders in Malawi. Gend Technol Dev 20(2):117–148

Ngigi MW, Müller U, Birner R (2018) Farmers' intrinsic values for adopting climate-smart practices in Kenya: empirical evidence from a means-end chain analysis. Clim Dev 10 (7):614–624

Nhamo L, Mabhaudhi T, Modi A (2019) Preparedness or repeated short-term relief aid? Building drought resilience through early warning in Southern Africa. Water SA 45(1):75–85

Nyasimi M, Huyer S (2017) Closing the gender gap in agriculture under climate change. Agric Dev 5(30):37–40

Nyasimi M, Kimeli P, Sayula G, Radeny M, Kinyangi J, Mungai C (2017) Adoption and dissemination pathways for climate-smart agriculture technologies and practices for climate-resilient livelihoods in Lushoto, Northeast Tanzania. Climate 5(3):1–22

Speranza CI, Wiesmann U, Rist S (2014) An indicator framework for assessing livelihood resilience in the context of social–ecological dynamics. Glob Environ Chang 28(9):109–119

Taylor M (2018) Climate-smart agriculture: what is it good for? J Peasant Stud 45(1):89–107

Teddlie C, Tashakkori A (2009) Foundations of mixed methods research: integrating quantitative and qualitative approaches in the social and behavioral sciences. Sage, Thousand Oaks

Walker B, Holling CS, Carpenter S, Kinzig A (2004) Resilience, adaptability and transformability in social–ecological systems. Ecol Soc 9(2):1–9

Weichselgartner J, Pigeon P (2015) The role of knowledge in disaster risk reduction. Int J Disaster Risk Sci 6(2):107–116

Williams TO, Mul ML, Cofie OO, Kinyangi J, Zougmoré RB, Wamukoya G, Nyasimi M, Mapfumo P, Speranza CI, Amwata D (2015) Climate smart agriculture in the african context. In: Feeding Africa: an action plan for African agricultural transformation. African Development Bank, Dakar

3

Greenhouse Gases Emissions in Agricultural Systems and Climate Change Effects in Sub-Saharan Africa

Winnie Ntinyari and Joseph P. Gweyi-Onyango

Contents

W. Ntinyari (✉) · J. P. Gweyi-Onyango
Department of Agricultural Science and Technology, Kenyatta University, Nairobi, Kenya
e-mail: gweyi.joseph@ku.ac.ke

Abstract

Climate change has been viewed to result from anthropogenic human activities that have significantly altered the Nitrogen (N) cycle and carbon cycles, increasing the risks of global warming and pollution. A key cause of global warming is the increase in greenhouse gas emissions including methane, nitrous oxide, and carbon among others. The context of this chapter is based on a comprehensive desktop review on published scientific papers on climate change, greenhouse emissions, agricultural fertilizer use, modeling and projections of greenhouse gases emissions. Interestingly, sub-Saharan Africa (SSA) has the least emissions of the greenhouses gases accounting for only 7% of the total world's emissions, implying that there is overall very little contribution yet it has the highest regional burden concerning climate change impacts. However, the values could be extremely higher than this due to lack of proper estimation and measurement tools in the region and therefore, caution needs to be taken early enough to avoid taking the trend currently experienced in developed nations. In SSA, agricultural production is the leading sector in emissions of N compound to the atmosphere followed by energy and transportation. The greatest challenge lies in the management of the two systems to ensure sufficiency in food production using more bioenergy hence less pollution. Integrating livestock and cropping systems is one strategy that can reduce methane emissions. Additionally, developing fertilizer use policy to improve management of fertilizer and organic manure have been potentially considered as effective in reducing the effects of agriculture activities on climate change and hence the main focus of the current chapter.

Keywords

Modeling · Pollution · Environment · Mitigation · Nitrous oxide · Carbon dioxide · Methane

Introduction

Globally, agriculture is the main contributor to greenhouse gases emissions (GHG) that is estimated to be between 10% and 20% of the total anthropogenic GHG emissions (Allen et al. 2020). From a baseline scenario, it is projected that GHG emissions from agriculture will be 1.7 gigatonnes by 2050. The GHG concentration especially CO_2 has increased by 40% since preindustrial times due to fossil fuels burning and also land-use changes. Current arable land cultivation practices have raised a great concern on the increase of GHG to the atmosphere. In sub-Saharan Africa, the effects are quite crucial but diverse as it has been associated with the variations of the season (rainy and dry) (Awazi and Tchamba 2019). As a result, this has significantly impacted the level of productivity, leading to food insecurity in this region. According to the US Environmental Protection Agency in 2010, the agricultural sector is the leading source of global GHG and accounts for 53% of carbon dioxide CO_2 emissions (Ronaghi et al. 2018). The distribution of CO_2 emissions is

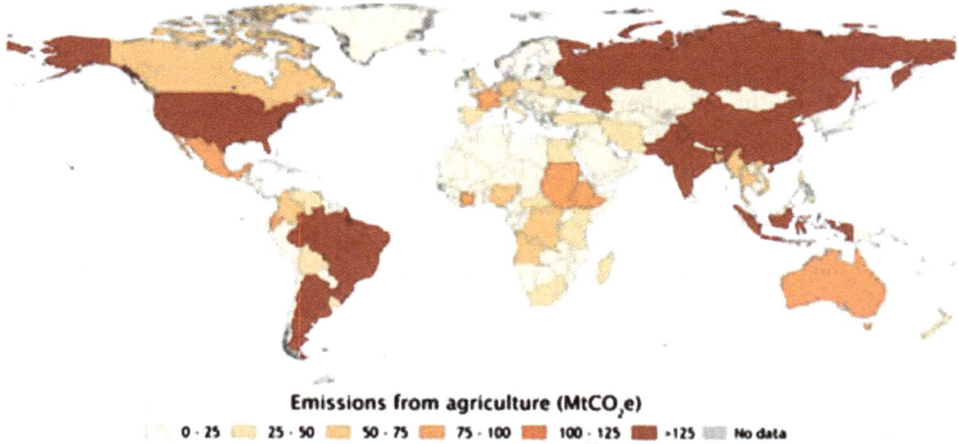

Fig. 1 Showing contribution of agriculture to GHG emission in global perspective. (Adapted from Malaka 2017)

globally spread as shown in Fig. 1, with most of sub-Saharan African countries emitting in the ranges of 0–75 $MtCO_2e$.

The main four sources of emissions from the agricultural sector include cropland soil management, rice cultivation, ruminant livestock enteric fermentation, and livestock manure management (Beach et al. 2015). According to Tongwane and Moeletsi (2018), half of Africa's total agricultural emissions were from enteric fermentation in 214 which signifies how great this source is. The GHG emissions in sub-Saharan Africa are estimated using Intergovernmental Panel on Climate Change guidelines in various subsystems of agriculture including crop production, animal manures, and methane emissions from livestock.

The main GHG fluxes from agricultural land include nitrous oxide (N_2O), carbon dioxide (CO_2), and methane (CH_4). The emissions of these gases (CH_4) and (N_2O) in agricultural land have increased by 17% from 1990 to 2005 globally (Popp et al. 2010). It is also anticipated that the GHG emissions will increase soon due to modern land-use changes that the sub-Saharan Africa region has adopted (Popp et al. 2010). Although the fluxes are anticipated to be high, there are various uncertainties in the estimates provided due to large insufficient temporal and spatial representation of emissions from agricultural soils (Mwanake et al. 2019). In SSA, these changes have significantly impacted the climatic patterns leading to more extended droughts and excessive rainfall that does not support agricultural productivity at the field level. Furthermore, organic soils are important sources of GHG emissions with about 35 thousand Giga grams of CO_2 as of 2010 demonstrated in SSA (Malaka 2017). Though there is no clear quantification data, there no doubt that there are some emissions from due to lack of management of crop residues since most of the agricultural waste in the land is openly burned or left free for livestock grazing. The burning of savannas and grassland fires that is common with the sub-Saharan African region is another greater contributors to GHG emissions (Malaka 2017). Due to this huge contribution of GHG emissions to the ecosystems, agriculture becomes an integral part that SSA and the whole globe should focus on to stabilize the emissions and save the current and future generations from the devastating effects of climate change. Therefore, this chapter will explore several available options and strategies that have the potential of mitigating GHG emissions from agriculture and minimize the impact of climate change.

The Approach Used

To acquire information of this chapter, we used the Google Scholar search engine to identify relevant peer-reviewed research articles published in high impact journals. A total of 53 articles were used that ranged from experimental, reviews, and policy papers. Keywords used in Google Scholar search engine were sub-Saharan Africa, GHG, emissions, management of emissions from the agriculture sector, climate change, agricultural sectors, management, agricultural systems, modeling, emissions, and global warming. The chapter focused broadly on various methods that have been suggested by the global panel, policies, and guidelines of the Intergovernmental Panel on climate change.

Key Strategies for Reducing Greenhouse Gases Emissions and Mitigation of Climate Change

Adopting Improved Nutrient Management in the Agricultural Lands

In sub-Saharan Africa, there is a problem of declining nutrient levels in cultivatable land which is due to poor nutrient management among the farmers. Most of the farmers lack the know-how on efficient use of the available fertilizers with minimal N loss but with improved crop yield (Galloway et al. 2003). To achieve this, farmers should be enlightened through policies and extension on various ways of fertilizer use that minimize the extent of nutrient loss to the environment that contribute to emissions of gases such as nitrous oxide (Huang et al. 2019). Improvement of nutrient use efficiency has a high potential of minimizing the N_2O emissions from cropland that are generally generated by the soil microbes from the surplus of N and indirect emissions from carbon dioxide from fertilizer companies (Galloway et al. 2003). Some of the recommended practices that farmers can adopt lie in the 4 R principle, which entails the use of the right source (fertilizer with higher efficiency), right rate, right timing, and right placement as expounded by Cassman et al. (2003). The proper placements of these fertilizers lead to fewer losses since there is a concomitant improvement in the plant uptake through making nutrients more accessible by the roots hence restricting available pathways for emissions. An important viable option for reducing losses is by reducing nitrification. Use of nitrification inhibitors for the case of nitrogen use is a promising strategy that has the huge potential of slowing the release processes that contribute to the formation of N_2O emissions (Coskun et al. 2017). Use of improved crop varieties with higher N uptake and use efficiencies will lead to a reduction in the nutrient loss since such crop idiotypes fully utilize the applied and available nutrient for use hence less losses in the form of GHG emissions (Sanz-Cobena et al. 2017).

Improved nutrient management is considered a better and viable strategy to overcome the challenges associated with the impacts of climate change in crop and food insecurities. According to García-Marco et al. (2016), adopting improved agronomic practices can lead to higher yields and promote the generation of more carbon that can be used to increase the soil carbon storage hence fewer losses to the environment. Some of the practices suggested by García-Marco et al. (2016) are extending crop rotations, use of improved varieties as mentioned earlier in this text, and using mixed cropping with perennial crops that will result to more carbon storage underground.

Emissions of GHG can also be reduced through practicing intensive cropping systems that minimize overreliance of pesticides and other agricultural inputs and consequently reduce the extent of emissions to the environment (Hoekman and Broch 2018). Besides, farmers/growers can use the cover crops that provide additional carbon to the soil and may help in extracting the available unused N by the next crops hence resulting to reduction of N_2O emissions (Oberthür et al. 2019). Use of Global Positioning Systems (GPS) guidance and variable rate technology are useful in applying inputs and hence allowing farmers to optimize nutrients with consequent less GHG emissions. In addition, the use of slow-release fertilizers such as prilled urea can be an efficient way of reducing N_2O emissions from cropland. Moreover, the use of biofertilizer has been advocated as an alternative to the more soluble and more reactive N sources as documented by excellent reviews by Ntinyari and Gweyi-Onyango (2018). Locally available nitrification inhibitors can also be useful in reducing the amount GHG from applied fertilizers to cropping systems. Soil pH management is also critical towards the management of N_2O emissions from the soil. Furthermore, the influence of pH on N_2O emissions more profound in soils that have high nitrate levels. According to the existing literature, pH contributes between 3 and 10 folds in incidences denitrification (Šimek and Cooper 2002). However, the good news is that use of liming in management of pH is something feasible among many farmers within sub-Saharan Africa if given fertilizer subsidies. Liming is a good option since is technically feasible and is affordable at a small scale levels. Nutrient use models are other useful tools in quantifying emissions from respective farms through the nutrient budget methods. These tools are key in estimating farm emissions by considering key components of nutrients inputs and nutrient outputs in given farms. In this case, it can be used to make a rough estimate of the amount of inputs required for specific crops to help in the reduction of emissions. Another option is the use and application of the soil systems budgets which record all the nutrient transformation related to emissions, for example, leaching, denitrification, and ammonia volatilization. This points to the fact that modeling tools are essential in linking the soil nutrients as well as a proxy resource for nutrient management and economic benefits to farmers.

Organic soils also need to be properly managed as they have the highest accumulation of carbon over the years. Avoiding row crops and tubers and deep ploughing using heavy machinery has been pointed out as one of the strategies that reduce the emissions of N_2O and CO_2 (Smith et al. 2008). In sub-Saharan

Africa, a large fraction of the arable land has been degraded due to soil erosion, loss of organic matter, acidification, and other processes. These have to a greater extent affected the potential of the soil in holding/storing of carbon (Smith et al. 2008). Therefore, there is need for reclaiming land by application of organic substrates like manures, biosolids, reducing tillage, and leaving crop residues in the soils. With these practices in the farm, high nitrogen use efficiency is anticipated and this will minimize the emissions of N_2O.

Proper Management of Residue and Adoption of Appropriate Tillage Methods

Advancing methods of weed control and machinery used in farming systems is another available way of reducing N_2O emissions from the soils. This is explained by the fact that soil disturbances tend to stimulate the soil carbon loss through enhanced erosion that result in soil carbon loss. However, the tillage method is dependent on the climatic conditions of the soils, considering that some reduced tillage options may have great effects of N_2O emissions from cropland (Feng et al. 2018). In tillage systems where growers/farmers retain crop residues, they tend to increase the soil carb since these residues act as precursors of soil organic matter which is a main store of carbon in soils. Mostly in the case where paddy crop growing is practiced, it has been found that no-tillage has a significant decrease in methane CH_4 emissions. Furthermore, recycling of crop residues has been reported to be of help in preventing the release and/or deposition of aerosols and GHGs that are generated during burning (Feng et al. 2018). Biomass burning is always thought to be a key contributory factor to climate change since it releases methane in notable amounts. This is a common practice in Kano plains in Western Kenya under rice production. This implies that there is a need to manage fire concerning biomass burning in agricultural fields. Reducing biomass burning will also go a long way in minimizing the emissions of hydrocarbons and reactive nitrogen emissions that react to form tropospheric ozone (Leng et al. 2019). This is explained by the fact that smoke composes of aerosols which can have either warming or cooling effects of the atmosphere, hence directly or indirectly contributing to climate change. In this light, reducing the frequency of or the intensity of fires will contribute to landscape carbon density in soil and biomass. To minimize the emissions, mitigation of the radiant forcing is expected to entail fire suppression strategies such as reducing fuel load through vegetation management and also burning of biomass at the time of year when CH_4 and N_2O are less emitted (Leng et al. 2019).

Crop rotation is advocated since it has given beneficial outcome through improving the quality of soils, with reported overall increment of as much 50% in terms of organic carbon into soils (Paustian et al. 2019). This practice creates resilience in cropping systems and helps the world to be able to combat climate change effects and hence can be a better option for SSA, where this is practiced already in noticeable scale. Through this approach, it is possible to sequester carbon for long-term storage of the atmospheric carbon. In this regard, this alternative offers the best solution to counter the greenhouse gases emissions in the ecosystem. In addition, more yields are obtained from the cropping systems, which helps reduce the direct impacts of climate-related contributions to food insecurity. In soil management, there is also a need to bring in low-cost inhibitors that regulate nitrogen processes in the soils. However, this requires knowledge on the sources of the GHG using various soil microbial process that can help to bring mitigation of the same on board. Diversifying on the crop rotations will also provide another strategy to minimize GHG emissions from agriculture in SSA. Diversification of crops helps in improving the productivity in the different and diverse agro-ecosystems in SSA and also lower the carbon footprint. The choice of these crops can be guided depending on the ecological requirements in various growing zones within the region. Another option in minimizing emissions is through intensifying rotations in cropping systems through fallowing. Fallowing promotes nitrogen mineralization hence increasing available nitrogen for use by crops and minimize the amount of organic matter hence less carbon is stored in the soils. Most of the farmers prefer burning of various crop residues in their farms which contributes to climate change in various ways. It contributes to releases of GHGs like CH_4 and also generates hydrocarbon and reactive nitrogen emissions that react to for tropospheric zones. Burning also destroys existing grass and also blackens the soil spoiling its quality and ability to sequester carbon. As a mitigation practice, farmers should rescue the frequency and the extent of fires for less CH_4 and N_2O emissions (Smith et al. 2008).

Introduction of Carbon-Sequestrating Grass Species

Introduction of improved grasses that have a high level of production or with adaptations to C allocation to deeper roots has a great potential of increasing the soil carbon (Yang et al. 2019). For instance, using deep-rooted grasses in savannas has been associated with increased rates of carbon accrual hence less is emitted to the atmosphere. Cultivating of legumes into grazing lands has also been associated with the promotion of soil C storage and perhaps reduces N_2O emissions (Garnett et al. 2017). Grasses also act as cover crops that have long-term potential for reducing GHG emissions through agriculture that comes from agricultural activities. The grasses have the potential of absorbing and retaining stored carbon in the soil. The roots and shoots of cover crops feed bacteria, fungi, and earthworms and other soil organisms that have a significant contribution to the soil carbon levels over time. In the land that is left fallow, there is a need to convert the lands to grassland to sequester more carbon and create a balance of the carbon in the soils for a longer period. The most beneficial aspects of grassland are that they can stay longer without ploughing hence reducing emissions of N_2O from the soils. In grasslands, intensity grazing has an influence on the density of the grasses and allocation of carbon into the grass fields. It has been reported that carbon accrual on optimally grazed land has been often greater compared to ungrazed grasslands. Also, irrigation of grasslands has also high promotion of the soil carbon gains. Also alleviating deficiencies of nutrients in grassland through use of fertilizer or other organic amendments will promote carbon soil storage (Smith et al. 2008).

Livestock and Manure Management Strategies in SSA

Livestock, specifically ruminants such as cattle and sheep, are important sources of CH_4 estimated to 18% of the global anthropogenic of this gas. In Table 1, there is a high chance of methane emissions that are more prevalent in the year 2000 compared with those likely to be there in the year 2030. These percentages represent a great share that needs to be mitigated using possible strategies that are available to the local farmers (Swamy and Bhattacharya 2006). According to the current statistics, the current population of livestock is anticipated to increase in 2030 based on the projections with sub-Saharan Africa, leading in the population as demonstrated in Fig. 2 (Herrero et al. 2008). This means that there is a need to come up with strategies for minimizing CH_4 from livestock systems. Some of the mitigation practices that can be used to reduce CH_4 emissions from livestock rearing include but not limited to improving the feeding practices. Feeding animals with more concentrates other than forages has been reported to reduce the rate of.

CH_4 for reducing. Other practices that can be used to minimize the level of CH_4 emissions are adding oils to the diet, as this is key in improving pasture quality since it improved animal productivity that reduces the proportion of energy lost as CH_4. A high intake of protein feeds is also associated with reduced N_2O emissions to the environment.

Another strategy of managing livestock methane emissions is the use of specific agents and dietary additives that can have the ability to suppress methanogenesis process. Some of the additives include Ionophores antibiotics that can minimize methane emissions in the livestock systems (Gibson 2002). Halogenated compounds inhibit methanogenic bacteria and their effects although they can have reduced feed intake. These have not yet been used extensively in SSA and have a huge potential of adopting. The bovine somatotropin (bST) and hormonal growth implants do not specifically suppress CH_4 formation, but by improving animal performance they can reduce emissions per kilogram of animal product (Cerri 2010), which is also another viable option. Adoption of longer-term management changes and animal breeding will enhance reduction of methane output per kilogram of an animal. With improved animal production efficiency, there are reduced lifetime emissions (Wall et al. 2010). Supplements can also be used in reducing methane emission in the livestock. Some of the supplements include oils, fats, tannins, probiotics, and marine algae. It has been reported that methane abetment between 10% and 25% is possible when feeding ruminants dietary oils. Also, plant secondary compounds like condensed tannins have proved to reduce methane production by 13–16% through curbing the process of methanogens.

Other available methane suppressers that are used in combating protozoal infections. Specialized proteins targeting methane-producing microbes will also constitute an effective alternative towards the management of emission from livestock. Moreover, plant saponins that occur naturally in many of the existing plant families have the potential to reduce methane and can be effective in the current measure towards minimizing the concentration of GHG in the atmosphere. Enteric methane emissions can be reduced through manipulation of the microbial communities in the

Table 1 Methane emissions by domestic ruminants in different livestock production systems in Africa 2000–2030. (Adapted from Novak and Fiorelli 2010)

System[a]	2000			2030					Differences 2000–2030						
	Cattle[b]	Goats[b]	Sheep[b]	Total[b]	Methane/km[2b]	Cattle[b]	Goats[b]	Sheep[b]	Total[b]	Methane/km[2c]	Cattle[c]	Goats[c]	Sheep[c]	Total[c]	Methame km[2c]
Livestock grazing															
Arid	1704.6	211.8	290.5	2206.9	123.6	2187.4	266.2	327.9	2781.5	157.3	0.28	0.26	0.13	0.26	0.27
Humid	217.7	24.6	14.6	256.9	179.5	261.1	29.4	14.9	305.4	318.8	0.20	0.20	0.02	0.19	0.78
Temperate	175.1	9.1	28.5	201.2	772.8	71.6	5.8	18.3	95.7	842.6	−0.59	−0.36	−0.36	−0.52	0.09
Total	2097.4	245.4	333.6	2665.0	136.3	2520.1	301.4	361.1	3182.7	169.7	0.20	0.23	0.08	0.19	0.24
Mixed															
Arid	2190.9	164.2	166.3	2521.5	650.6	4273.0	328.1	292.3	4893.4	986.5	0.95	1.00	0.76	0.94	0.52
Humid	494.2	71.9	38.9	605.1	427.5	804.0	160.8	71.0	1035.8	587.8	0.63	1.24	0.82	0.71	0.37
Temperate	1120.2	36.6	56.4	1213.2	1932.3	885.2	26.8	42.2	954.2	2958.2	−0.21	−0.27	−0.25	−0.21	0.53
Total	3805.3	272.8	261.7	4339.8	733.2	5962.2	515.7	405.5	6883.4	977.1	0.57	0.89	0.55	0.59	0.33
Other	451.0	49.0	41.9	541.9	135.3	614.6	79.3	62.7	756.7	203.5	0.36	0.62	0.50	0.40	0.50
Total methane from enteric fermentation	6353.7	567.3	637.1	7546.7	256.0	9097.0	896.5	829.3	10822.8	366.7	0.43	0.58	0.30	0.43	0.43
Total methane from manure	190.6	17.0	19.1	226.4	7.7	272.9	26.9	24.9	324.7	11.0	0.43	0.58	0.30	0.43	0.43
Total methane emissions for African domestic ruminants	6544.4	584.3	656.2	7773.1	263.7	9369.9	923.4	854.2	11147.5	377.7	0.43	0.58	0.30	0.43	0.43

(continued)

Table 1 (continued)

	2000			2030					Differences 2000–2030						
System[a]	Cattle[b]	Goats[b]	Sheep[b]	Total[b]	Methane/ km^{2b}	Cattle[b]	Goats[b]	Sheep[b]	Total[b]	Methane/ km^{2c}	Cattle[c]	Goats[c]	Sheep[c]	Total[c]	Methame km^{2c}
By environment															
Arid	3895.5	376.1	456.8	4728.4	217.6	6460.5	594.3	620.2	7675.0	339.0	0.66	0.58	0.36	0.62	0.56
Humid	711.9	96.5	96.5	861.9	302.9	1065.2	190.2	85.9	1341.2	493.1	0.50	0.97	0.60	0.56	0.63
Temperate	1295.3	45.7	84.9	1414.4	1592.4	956.7	32.6	60.6	1049.9	2407.1	−0.26	−0.29	−0.29	−0.26	0.51
Other	451.0	49.0	41.9	541.9	135.3	614.6	79.3	62.7	756.7	203.5	0.36	0.62	0.50	0.40	0.50

[a]Methane from enteric fermentation
[b]In million kg
[c]As a proportion of 2000 values

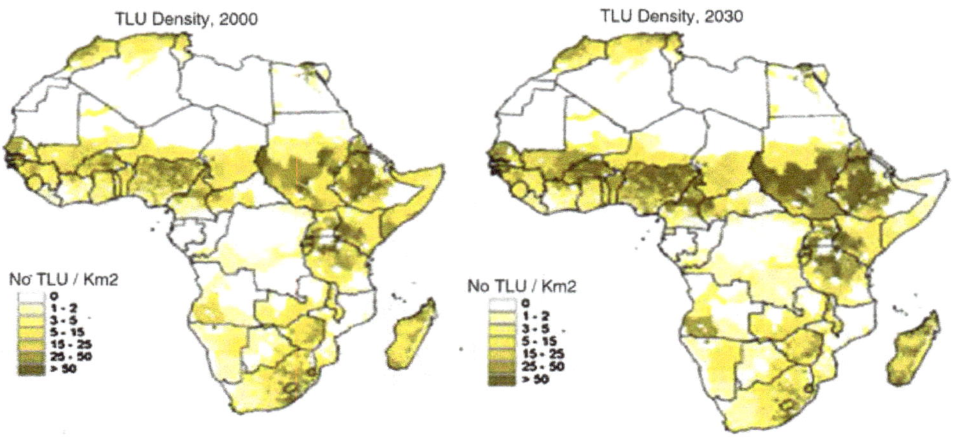

Fig. 2 Showing spatial distribution of livestock in Africa in the year 2000 and projected for 2030 (Herrero et al. 2008)

stomach of ruminants. This can be done through injecting/administering specific vaccines, drugs, or supplements that contribute to the reduction of methane. Other methods inlcude; biological methods can be adopted to reduce the ability of the livestock to reduce methane production in domesticated animals. For instance, the virus that attacks microbes in the rumen of the ruminants can be used to ensure animals produce less GHG. Methanotrophs which are microbes can also be introduced in the animals to help in the breakdown of produced into other wastes. The bovine somatotropin and hormonal growth implants is also a possible option but it does not suppress methane formation but rather improved the general performance of animals hence minimizing the intensity of GHG.

Manure management should be given priority as it is associated with significant amounts of N_2O and CH_4 during storage. However, the magnitude of the emissions varies depending on the structure. World over, there are no well-defined measures and policies of manure use and management since most of the grazing is done in open fields. For manure stored in lagoons or tanks, methane emissions can be minimized by covering or cooling of the sources. Other preliminary information has also suggested that covering manure heaps can significantly reduce N_2O emissions (Dennehy et al. 2017). In addition, emissions from manure can be curtailed by altering feeding practices and also precision application when it comes to cropland. Methane digesters/anaerobic digester can also be used as an alternative strategy to the reduction of GHG although it is expensive to install. This is since maintaining anaerobic digester requires basic knowledge in wastewater and electrical generation hence high investment costs that can be managed by respective governments. Adopting anaerobic digestion offers an excellent strategy towards the reduction of the CH_4 and CO_2 to the ecosystem. The process offers natural existing bacteria that decompose organic matter resulting in biogas (Wang et al. 2017). In return, the biogas can be used to burn fuel and reduce emissions of GHG from fossils. Other practices that need to be put into consideration include:

- Application placement (e.g., slurry injection)
- Application timing
- Application amounts (e.g., controlled rate systems)
- Export of manure (from the agricultural system)

GHG models can be developed at whole-farm level approaches to help in mitigation of various emissions specifically from daily farms. The advantage of using models in dairy farms is their ability to simulate calculations of CH_4 and N_2O emissions although the models may considerably vary but they give information on the current and future trends upon application of given measures.

Following the Intergovernmental Panel on Climate Change standard emission factors, it is confirmed that nitrogen deposited from urine and feces is converted to nitrous oxide twice the rate from the fertilizers used in farms. From the projections done for 2050, it is shown clearly that that in open pasture method, if no mitigation is put into practice, there is a possibility of having 25% increment on the emissions. In open field grazing strategy, the farmers can use chemical nitrification inhibitors as a way of preventing the transformation of the nitrogen is excreted in urine and feces to nitrous oxide. The other option is the introduction of biological nitrification, for example, use of the Brachiaria grass that has been reported to generate almost zero N_2O emissions. Therefore, this a grass that the breeders can focus on to have the particular trait transferred in mostly grown grasses in pastures fields (Fig. 3).

Use of methane digesters is another excellent successful strategy for consideration in SSA but will require government intervention especially in extending facilities from e large to small scale farmers to be able to encourage innovation. The respective governments should come into a collaboration to develop the most cost-effective technologies and this can be achieved through funded programs by innovators. In addition, there is a need to have programs that will enable early detection and remediation of leakages from digesters. The collected manure can also be used to make biogas which provides an additional source of renewable energy hence has higher chances of reducing GHG emissions. This is an economical method since it will save the farmers from using fossil fuels in cooking since they can generate power from their farms. It is also feasible to acidify slurry produced mostly in the zero-grazing sectors to enhance to reduce conversion and formation of methane.

Adopting Bioenergy as GHG Emission Reduction Strategy

Crops and residues have high potential as sources of feedstocks for energy as a replacement of fossil fuels. Many of the materials that have been proposed for bioenergy role are grain, crop residue, cellulosic crops, and various tree species (Haberl et al. 2012). The advantage of these products is that they can be burned directly but later processed to generate fuel liquid. Although this process also produces CO_2, it is associated with the original CO_2 from the atmosphere for the process of photosynthesis hence displaces the CO_2 that could be produced from fossils. The

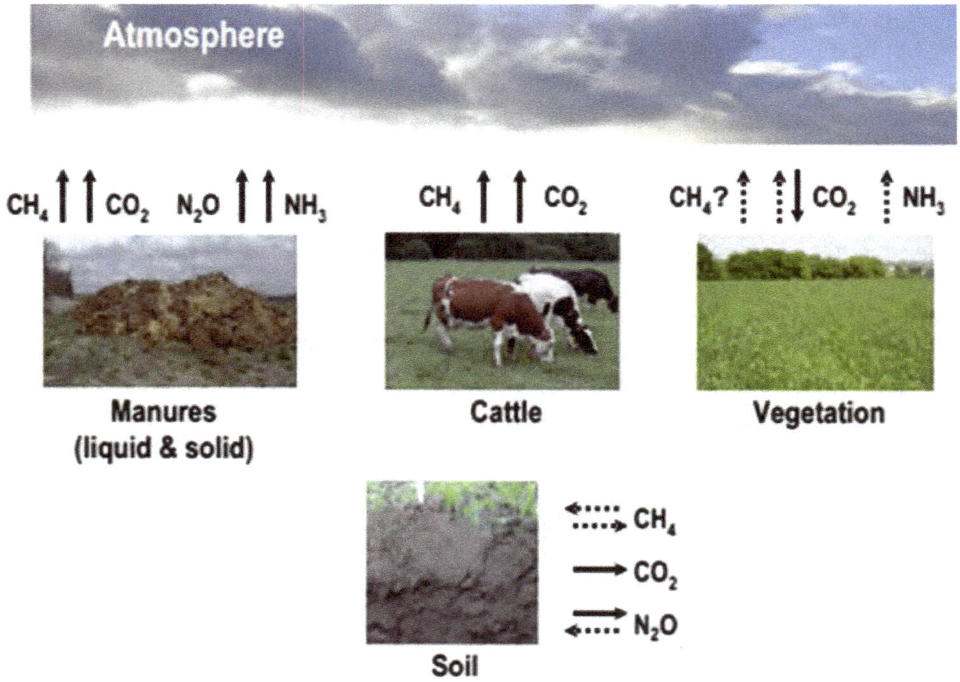

Fig. 3 Greenhouse gas and ammonia fluxes in the main compartments of a mixed crop-dairy system. (Adapted from Novak and Fiorelli 2010)

energy used in growing and processing the feedstock of bioenergy has a net benefit to the atmospheric CO_2 (Eloka-Eboka et al. 2019). The existing interaction of an expanding bioenergy sector land use and other ecosystems services such as food production has not been exploited potentially; for instance, through assessment modeling, sub-Saharan Africa has been listed as one of the promising regions for bioenergy. It is estimated that in 2050, 65% of the expected agricultural energy emissions will arise from farm energy use. In this regard, there is a need to reduce the amount of farm energy used from fossil to nonfossil use. Mitigation on the use of diesel fuels by tractors and other heavy equipment will be an effective move but requires transitions to hydrogen power that comes from solar and wind power. Also, battery-powered equipment and synthetic carbon-based fields will provide an alternative strategy. Use of these kinds of renewable energy in farms where mechanization is highly practiced will help in mitigating 85% of the total emissions from the synthesis of nitrogen fertilizers. Considering bioenergy as an option of mitigating GHG and saving our planet from climate change impacts, there is a need to make major transitions that are essential in exploiting the great potential of this particular source of energy (Smith et al. 2008). Bioenergy or use of biofuels implies a decrease in the dependence on fossil fuels and chemicals. The energy supplied from the use of plant-based fuels has the potential of decreasing GHG flow to the atmospheres. The biofuels will enable GHG mitigation through three key factors that include lower and competitive prices compared to fossil energy sources, ease in their production since the plants are grown during the entire year, and ecological and economic benefits of biofuels due to their environmentally friendly nature. Through these strategies, it will be efficient in mitigation of climate change effects in sub-Saharan Africa, though this will take longer to be achieved compared to other considered options.

Reduction of GHG Through Genetic Selection of Animals

Dairy is being one of the highest emitters of GHG to the atmosphere that requires critical strategies right from the selection of the best animals to reduce methane production. More focus should be put forward on the selection of cows that have less production of CH4 per unit according to Table 2. The impact of increasing genetic potential of the production stock is to increase the daily live weight gain of livestock for meat production or by nutrition hence leading to reduction of methane product per unity of the product. However, faster growth for slaughter implies that more feed will be needed to meet the required weight within a short time (de Haas et al. 2017). Therefore, this also means that the genetic selection has also to consider the type of grasses that can meet the requirements of such animals within the set time.

A high growth genotype leads to low emissions hence providing a promising measure towards mitigation of the climate change as a result of concentration in the atmosphere (Martinez-Fernandez et al. 2018). In addition, the use of environmentally fit genotypes is fundamental to increasing efficiency of converting feed to beef. In an experiment conducted in a farm in Northern Territory, methane production was lower by 31% in per tonne of the weaned weight within a given time because of a genetic characteristic of animals that can be weaned early to reduce the total emissions in the livestock sector (Mottet et al. 2017).

Adopting to Isotopic Tracers in the Agricultural Field

Stable isotopes that do not emit radiation such as nitrogen-15 and carbon 13 on small experimental fields are good options. This method allows then to analyze the efficiency of the crops to consume nitrogen and the accumulation of carbon in the soils. For instance, nitrogen-15 technique is widely used by scientist to track and advise farmers on the amount of chemical fertilizer/manure needed by the plants as one of the ways to reduce increased emissions to the ecosystems (Slaets et al. 2016). Nitrogen-15 isotopic techniques can also be used by scientist to identify the sources of nitrous oxide in production and this forms an important pathway of reducing the emission of the gas to the atmosphere. Therefore, this technique has the potential of measuring the impact of climate and provides a way of mitigation.

Carbon-13 has also been used in the assessment of the soil quality by identifying the content of organic sources of carbon in the soil. This is also a crucial favor in promoting the optimal application of the agricultural practices hence reducing GHG emission. Stable isotope can also be essential in the generation of information on

Table 2 Feeding strategy and genetic selection in the management of GHG emissions. (Adapted from Novak and Fiorelli 2010)

Mitigation options for livestock management		CH_4	N_2O	CO_2	NH_3
Feeding strategy	Adding linseed lipids to the diet	↘?	–	–	–
	Increasing the proportion of concentrate in the diet	↘from animals ↗ from slurry?	↗ or ↘? (from slurry[a])	↗ (fossil energy + soil)	↗ or ↘? (from slurry[a])
	Increasing the proportion of maize silage in the diet	↘ from animals	–	–	–
	Introducing legumes into grazed grasslands	↘	↗?	↘?	↗?
	Limiting excess N in the diet	↗?	↘	–	↘
Genetic selection	Selecting cows with low enteric	↘?	–	–	–
	CH_4 production				↘[b] or ↗?
	Selecting high-yielding cows	↘ or ↗?	–	–	–
Herd characteristics	Reducing the replacement rate	↘?	–	–	↘
	Reducing the number of milking cows	↘	↘	↘	

[a]Depending on the N content of the concentrate compared to the roughage
[b]Results from Lovett et al. (2006)
↘: the mitigation option decrease the emissions
↗: the mitigation option increase the emissions
"↘ or ↗": both tendencies have been shown
–: no information was given on this compound
0: studies have shown that this option had no significant effect on this compound.
?: the result needs to be confirmed by more studies.

GHG production and transport within a soil ecosystem. Importantly, stable isotopes have the potential of studying how GHG exchange with soils and atmosphere. This information is useful to enable researchers to understand the pathways of GHG, as well as the magnitude, as well as the magnitude is hence given rational recommendations (Zhu et al. 2019). However, this method is expensive and local farmers cannot afford to use to make various estimates. Therefore, the use of experimental reference sites can be useful in giving recommendations over a geographical position with similar activities.

Management of Emissions in Rice by Adoption of Efficient Varieties

Rice production in flooded or paddy rice contributes to at least 10% of global agriculture production. There are various suggestions that have high technical potential to mitigate rice emissions and that many of the options have a high potential of giving economic gains through yields and reduce the amount of water used in irrigation. Focusing on yield increase will give a promising measure towards reducing emission in rice production per unit area (Saha et al. 2018). This is in agreement with the FAO's forecast that increasing the rate of yield in paddy will also contribute to emission reduction in rice. Removing rice straws from paddies before reflooding to reduce methane production is also advised as one of the practices that farmers can embrace. The straws may be used for Bioenergy production (Joint 2018). Reducing the rate of flooding will reduce methane-producing bacteria where farmers can draw water down during the middle of the growing seasons. Another option is the beeding for rice varaities with lower emmissions of methane in the systems. Experiments in other regions such as China and Japan have shown that drawdown can reduce methane emissions by up to 90% (Müller et al. 2016). Use of the system of Rice Intensification that aims at reducing irrigation water at farm-scale is also an adaptable technique that farmers can use to minimize methane emissions in their rice fields. In rice irrigation, farmers should be advised to reduce the duration of flooding to minimize the populations of the methane-producing bacteria. In this regard, rice can be planted in a dry field other than flooding them and also they can reduce water during the middle days of the growing season. Breeding of lower methane rice will be another strategy that researchers should focus on to introduce both upland and lowland rice-growing irrigation systems. Water management in irrigation through the expanding area on irrigation could be more effective in increased carbon storage in the soils. Draining of agricultural fields in humid regions will help in suppressing N_2O emission through improved aeration in soils. This strategy has been used in other regions such as China and Japan since it also increases yield and saves on irrigation water hence can also be adopted for the sub-Saharan Africa region.

Development of Flexible Technology-Forcing Regulations

The GHG emissions are strong issues that need to be regulated by the governments more so in sub-Saharan Africa counties where the governments have not invested much on controlling emissions due to their low-income capabilities. This is since if this issue remains a voluntary service, most of the individuals are not likely to take it as seriously as it is taken globally (Inglesi-Lotz and Dogan 2018). Therefore, there is a need to have flexible regulations that are designed to spur the technological development for the needed change. For instance, in case of fertilizer use, countries should develop regulatory systems that are similar to those developed in the United States or other places to increase the fuel efficiency of fleets of time (Nyamoga and Solberg 2019). Also, fertilizer manufacturers and importers will be required to sell fertilizers with increased efficiencies. Fertilizers regulation needed to be brought into consideration, for instance, the manufacturer should be barred from releasing any fertilizer product that does not have a coat to enhance its slow release and minimize N_2O emissions in cropping systems.

Minimizing Enteric Fermentation and Food Wastes

Enteric fermentation has the highest accountability of emissions from agriculture. It is part of the digestive systems of herbivorous animals that have a large four-compartment stomach with a complex microbial environment that enables it to digest complex carbohydrates. The process of digestion produced methane as a byproduct hence contributing to its accumulation in the atmosphere. In this regard when animals are kept for long, the more they produce methane, and therefore, this should be shortened and ensure the levels of methane per production unit are minimal (Owen and Silver 2015). The three main existing options to mitigate this kind of fermentation and emissions are

1. Improving the quality and digestibility of feed
2. Providing supplements and additives and reduce methane
3. Optimizing the health and reproductive capacity of the herds

Food and Agriculture Organization (FAO) has approximated that food wastes contribute to almost a third of the world's GHG. This is more so in sub-Saharan Africa due to the culture of people holding a social gathering with several types of edibles available hence providing a wide range of choice. These foods are potential sources of GHG emissions with key gases produced being carbon dioxide, methane, nitrous oxide, and hydrofluorocarbons especially in countries without any existing policies on food loss and waste management (Capone et al. 2016). This calls for the development of a monitoring framework for developing countries especially in sub-Saharan countries through installing simple tools that can be used in the traceability of GHG emissions. According to the Intergovernmental Panel on Climate Change (IPCC), the lowering of food wastes will help in reducing the rate of GHG emissions and concentration in the environment. In this regard, there is a requirement for the respective government to come up with policies and campaign to enlighten the locals on the need to reduce their food wastes as one of the factors for environmental sustainability and to alleviate the serious threats of climate change that are currently experienced (Edenhofer et al. 2011).

Use of Biochar to Minimize GHG Emissions

Biochar is an organic material rich in carbon that is produced by the heating of biomass in the limited supply of oxygen. It is mainly produced as an additive to soils to improve nutrient retentions and carbon storage. In recent research, there has been growing interest in the use of biochar as an amendment to improve soil health, decrease net N_2O emissions and methane, and to store carbon in the soils. Biochar has been recognized as one of the possible materials to enhance reduction of CO2 concentration in the environment. The advantage of biochar is sufficiently used to reduce GHG emission in the atmosphere for a prolonged time. It also allows a net reduction on the emissions in the entire lifecycle including both direct and indirect land uses (Kumar et al. 2020. It has been documented that biochar has the potential of reducing anthropogenic CO2 in the atmosphere by 12%. Biochar has a low degradation process which helps in increasing its potential to store more carbon in the soils compared to other materials.

Compositing of Solid Manure Before Applying to Crop Fields

In many organic systems, composting of solid manure before applying to the field is encouraged. This method offers the advantage of producing a more stable product, free of weeds and toxins and easier to spread in the cropping systems. Through compositing it is subjects manure to the aerobic composition at temperatures of around 60 °C. Since this method requires frequent turning of the heaps, the method makes it more vulnerable for NH_3 losses; however, after application to the cropping fields, there are minimal losses since remaining N is mainly bound organically (Kumar et al. 2020). The mechanical turning of the composted manure is associated with lower N_2O and CH_4 compared to the anaerobic method; hence, this should be considered for the whole management systems right from manure storage to supplication to soils in the fields. It has been estimated that compositing lowers N_2O and CH_4 emissions; although significant amounts of CO_2 are emitted during the composting process, it is not considered as a net source of CO_2 in the entire agricultural chain. In addition, since composted manure is less degradable than fresh manure, application of this is a greater source of carbon storage in the soil and thus reduces the rate of CO_2 emissions. The rapid incorporation of manure in the soils as an improved technique of application is associated with fewer emissions and is more cost-effective (Joint 2018). This kind of incorporation should be done during the first hours after application. Through this method, a significant N_2O emissions in the soils will be reduced. Farmers should also consider using farmyard manure over liquid slurry for reduced GHG emissions.

In crop production as agriculture, subsystem offers various mitigation of greenhouse gases specifically the CO_2 and N_2O that are main gases in this level. The limitation of net CO_2 can be achieved through increasing the rate of carbon storage to soils or by the plant. This is achieved through slowing return of stored carbon into the atmosphere via mineralization (Stanton et al. 2018). Most importantly, N_2O mitigation should focus specifically on improving the nitrogen use efficiency of crops since N_2O is generated from the soils. In crop production, farmers should also avoid compacting during tillage since this may increase N_2O emission due to the conducive environment created for anaerobic zones within the soil structures. During the incorporation of plant, farmers should be advised to use plant contents with low N content since high N content in the tissue will trigger mineralization and denitrification that can alleviate the levels of N_2O emissions. Tackling the interlinked problems on land degradation, climate change mitigation is a promising strategy that can produce promising results for both farmers and government (Zerhusen et al. 2019).

Integrated Farming Systems

The integrated farming system is another way to reduce GHG emission and promotes the fight against climate change. The integrated systems are based on nuclear techniques that have been put into practice in some counties within sub-Saharan Africa including Kenya and Uganda. The practices are aiming at maximizing the recycling of nutrients that are found in animal manure and crop residues. This will ensure there is a reduction in the chemical fertilizers use and this will enhance the reduction of GHG from agricultural land (Stanton et al. 2018). This is a practical strategy since farmers can recycle nutrients as livestock mainly feeds on herbs and grasses that are excreted in form of manure then farmers can collect the manure and apply in fields, which is a way of returning many nutrients into the soil as illustrated in Fig. 4.

This strategy is feasible and practically possible in sub-Saharan Africa since it has been applied in other regions and does not require high investment but rather involves a set of regulations and extension to enlighten farmers on the benefits of

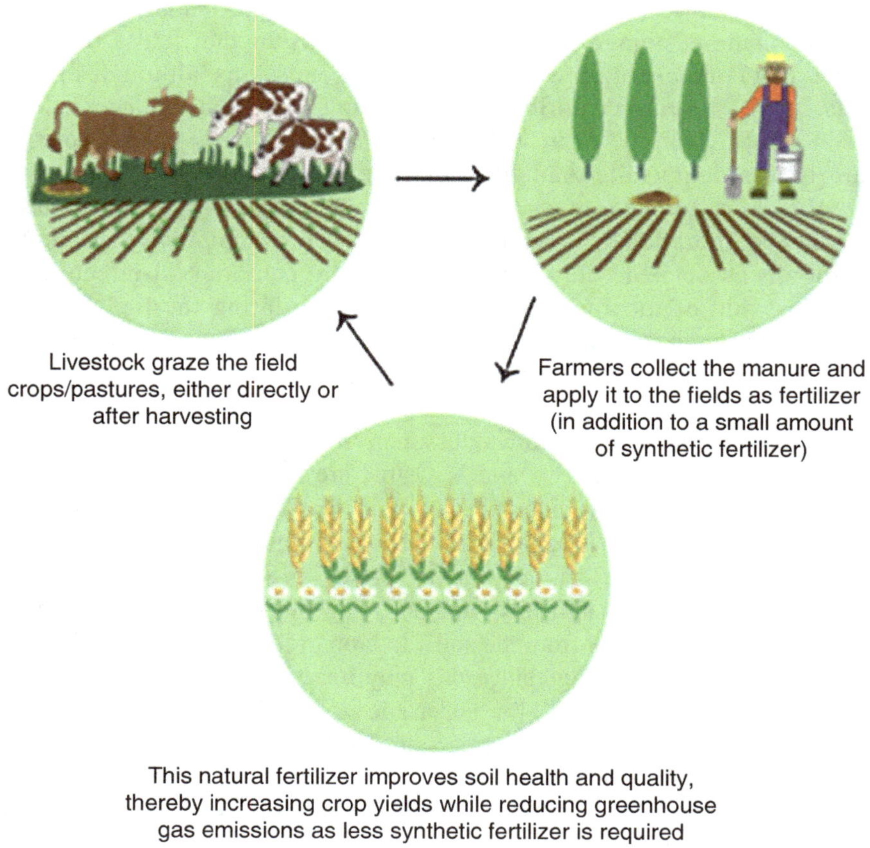

Fig. 4 Illustration of how an integrated system can reduce greenhouse gas emissions. (Adapted from Garnett 2011)

this process. Integrated cropping–livestock practices have been used in Brazil and have given a successful out management plan in the use of land more efficiently. As a result, the GHG emission from urine and dung was reported to reduce by 89% from respective farms. Polyculture is another agronomic practice that can be used to mitigate emissions of GHG and help in combating the adverse effects of climate change. Polyculture entails growing of multiple crops in the same space to increase local biodiversity and increase the soil carbon. Agroforestry is another agronomic practice that can be encouraged as a strategy to minimize GHG emission in agriculture. This practice entails integrating crops or livestock and trees either for timber, firewood, or any other wood products. The agroforestry provides buffer strips in the riparian land as well as increasing carbon stock in the soils since they increase carbon sequestration hence emissions of CO_2 to the atmosphere (Garnett 2011). Changes in consumption habits have substantial effects on the reduction of GHG emissions. Therefore, this calls for addressing how our food from agriculture is distributed across a supply chain. In particular, many of the environmental bodies have a recommended reduction of consumption of meat and other dairy products to enhance the reduction of GHG emissions from agricultural production. This implies that if consumers can adhere to this strategy and focus on well-grown crop products will have a significant effect on the overall concentration of GHG in the atmospheres. How this requires policy regulations and consumers' willingness to adapt to this kind of change. Although it is difficult to some communities that rear livestock as the main agricultural activity, they could be advised to minimize meat consumption, which may be from 5 to 2 days a week.

Barriers to Mitigation GHG Emission and Climate Change in Sub-Saharan Africa

Although there are various opportunities to combat GHG emissions and reduce impacts of climate in Africa, there are key barriers that may hinder adoption and achievement of these goals. There is a challenge on lack of awareness among the farmers on the bottom line impacts of climate change and causes that emanate from their cropping-livestock systems. A great population of the farmers in sub-Saharan Africa are not aware of the existing option, either high or low cost. This is since there is less focus on the ground level to address the issue; in this regard mitigation of GHG and reducing the impact of climate change more focus on starting ground level for the change to experience. Moreso, there is the existence of unsureness about the practicality of the available measure among the leaders and activist of climate change actions; hence, less awareness has been created to date. There is also the existence of complex interactions as farmers do not know the bottom line of the unintended consequences on the complexity of interaction during the adoption of what is believed to make a change to GHG emissions from agriculture (Mehra et al. 2018). Regulations and policies are also other barriers towards adoption and application of the existing measure. Some of the verification and regulation are quite expensive, and the fact that sub-Saharan Africa is a low-income region confirms the weight of the barrier. In addition, new regulatory regimes may require structuring the existing regulation before the implementation of proper governing policies.

Adopting some of the options may be against the social norms of the culture with diverse cultures being dominant in sub-Saharan Africa. For instance, manure management may be viewed as a dirty practice by some communities or religious beliefs hence making its management difficult. Besides, most farmers may fear loss aversion and may be adamant towards investing on best practices as advised by regulators. The existing extension services have the challenge of meeting farmers need and they create doubt on bits of advice they receive from them. Also, the government is slow in making clear on the role of extension officer and to what extent do they need to give communal seminars on critical issues such as climate change and GHG emissions (Allen et al. 2020). Development and transfer of technologies is another underlying barrier in the management of GHG emissions and mitigation of climate change. This is due to limited human and institutional capabilities to enhance proper dissemination of new information. Technological barriers exist in the limitation of generating and applying the developed solution to handle the problem. Technologies development and transfer remain crucial components for sustainable increase of productivity and mitigation of climate effects. Ecological barriers exist because mitigation potential in agriculture subsystems is mainly site-specific even within the same cropping systems. The limited resources in Sub-Saharan Africa pauses an ecological barrier towrads management of GHG and climate changes impacts. The frequent drought in Africa may limit the exact measurement and application of the mitigation measures at the farm level. In addition, due to high levels of poverty in the region, there is a likelihood that affording good equipment for mitigation will be hard hence have to rely on donors from other regions.

Need for More Research

Mitigation of GHG emissions is a critical area that currently requires a lot of investment in research, because these gases are complex and emitted from day to day activities that men perform to earn a daily living. Due to the adverse effects, they are causing damage to our ecosystems; identification of scientifically proven measures and option needs to be analyzed and scrutinized keenly through research. The researcher should focus more on agriculture although other sectors such as transport, energy, and industrialization have an equivalent measure. Although several measures have been documented to curb and minimize the rate of emissions, there is a need to have additional measures and strongly emphasize on their adoption through policy regulations. The research should focus on both generating new technologies and management systems with lower emission rates. Therefore, this requires the collaboration of governments in the sub-Saharan each to generate funds that can motivate and facilitate scientists in agricultural discipline to focus more on GHG about to actual quantification of the emissions as well as documenting available measures in the current context. Generation of more knowledge through research will provide cost-benefit analysis concepts to help in the assessment of the trade-off in the climate change mitigation options (Garnett 2011). The discoveries through research will enhance the provision of information and rationale on taking actions to mitigate GHG emissions. In addition, it will form a basis on informing decision-makers and policymakers in the whole system.

Conclusions

In the light of adverse climate change effects being experienced in sub-Saharan countries currently, several opportunities to mitigate GHG in agriculture exist and can be realized through overcoming several barriers. There is a need to assign mitigation measures on the key causes of climate change hence more emphasis on greenhouse gases emitted from agriculture. The key GHGs – CO_2, N_2O, and CH_4 – are not only produced in the farm setting but other sectors including energy, transport, and industries have a significant contribution. However, more potential lies in the agricultural sector that is widely spread in major regions of sub-Saharan Africa. Critical elements that require proper management are livestock systems and crop production that have shown to have the highest emission of the main GHG to the ecosystems. Therefore, to protect and improve the quality of our planet, there is a need to apply measures including manure management, improved fertilizer use efficiency, genetic selection of animals with less methane emission, minimizing enteric fermentation and floodwaters in agriculture and dairy subsystems. Developments of feasible and technological solutions to minimize GHG at small scale level, adopting bioenergy, growing of grass species with carbon sequestration effects, integrating crop-livestock system, and managing emission in paddy rice trough proper use of irrigation water and biochar application among others. Putting all measures discussed in the current chapter will help reduce the impacts of climate change in sub-Saharan and improve the quality of air and environment health in general. Regardless of several feasible options existing to mitigate GHG emissions and climate change in agriculture, to achieve these measures, there is a need for collaborative efforts between governments and the locals to carry out the required ground activities. There is the existence of certain barriers that limit the potential of management. Some of these barriers include policy and regulation, inadequate human and institutional capabilities, unstructured extension services, fear of loss aversion among farmers, and absence of platforms for disseminating new technologies. For any of the above-discussed measure, decision-makers need to make a consideration to the extent to which it moves away or closer to making the environment safer and still maintaining sustainable agricultural production with less GHG emissions.

References

Allen J, Pascual KS, Romasanta RR, Van Trinh M, Van Thach T, Van Hung N, Chivenge P (2020) Rice straw management effects on greenhouse gas emissions and mitigation options. In: Sustainable rice straw management. Springer, Cham, pp 145–159

Awazi NP, Tchamba MN (2019) Enhancing agricultural sustainability and productivity under changing climate conditions through improved agroforestry practices in smallholder farming systems in Sub-Saharan Africa. Afr J Agric Res 14(7):379–388

Beach RH, Creason J, Ohrel SB, Ragnauth S, Ogle S, Li C, Salas W (2015) Global mitigation potential and costs of reducing agricultural non-CO2 greenhouse gas emissions through 2030. J Integr Environ Sci 12(sup 1):87–105

Capone R, Adriana C, Lee TKW (2016) Food losses and waste: global overview from a Mediterranean perspective. Mediterra 2016. Zero Waste in the Mediterranean. Natural Resources, Food and Knowledge/International Centre for Advanced Mediterranean Agronomic Studies (CIHEAM) and Food and Agriculture Organization of the United Nations (FAO)–Paris: Presses de Sciences Po, 193

Cassman KG, Dobermann A, Walters DT, Yang H (2003) Meeting cereal demand while protecting natural resources and improving environmental quality. Ann Rev Environ Resource 28:315–358. https://doi.org/10.1146/annurev.energy.28.040202.122858

Cerri C (2010) Management practices for greenhouse gas emission reduction and carbon removal in Brazilian agriculture, livestock and forestry. Centro de Energia Nuclear na Agricultura/ Universidade de São Paulo, Piracicaba

Coskun D, Britto DT, Shi W, Kronzucker HJ (2017) Nitrogen transformations in modern agriculture and the role of biological nitrification inhibition. Nat Plants 3(6):1–10

de Haas Y, Pszczola M, Soyeurt H, Wall E, Lassen J (2017) Invited review: phenotypes to genetically reduce greenhouse gas emissions in dairying. J Dairy Sci 100(2):855–870

Dennehy C, Lawlor PG, Jiang Y, Gardiner GE, Xie S, Nghiem LD, Zhan X (2017) Greenhouse gas emissions from different pig manure management techniques: a critical analysis. Front Environ Sci Eng 11(3):11

Edenhofer O, Pichs-Madruga R, Sokona Y, Seyboth K, Kadner S, Zwickel T, Matschoss P (eds) (2011) Renewable energy sources and climate change mitigation: special report of the intergovernmental panel on climate change. Cambridge University Press, Cambridge, MA

Eloka-Eboka AC, Bwapwa JK, Maroa S (2019) Biomass for CO2 sequestration encyclopedia of renewable and sustainable materials; Elsevier Inc. All rights reserved. https://doi.org/10.1016/B978-0-12-803581-8.11029-X

Feng J, Li F, Zhou X, Xu C, Ji L, Chen Z, Fang F (2018) Impact of agronomy practices on the effects of reduced tillage systems on CH4 and N2O emissions from agricultural fields: a global meta-analysis. PLoS One 13(5):e0196703

Galloway JN, Aber JD, Erisman JW, Seitzinger SP, Howarth RW, Cowling EB, Cosby BJ (2003) The nitrogen cascade. Bioscience 53:341–356. https://doi.org/10.1641/0006-3568(2003)053[0341:TNC]2.0.CO;2

García-Marco S, Abalos D, Espejo R, Vallejo A, Mariscal-Sancho I (2016) No-tillage and liming reduce greenhouse gas emissions from poorly drained agricultural soils in Mediterranean regions. Sci Total Environ 566:512–520

Garnett T (2011) Where are the best opportunities for reducing greenhouse gas emissions in the food system (including the food chain)? Food Policy 36:S23–S32

Garnett T, Godde C, Muller A, Röös E, Smith P, De Boer IJM, Van Zanten HHE (2017) Grazed and confused?: ruminating on cattle, grazing systems, methane, nitrous oxide, the soil carbon sequestration question-and what it all means for greenhouse gas emissions. FCRN, Oxford

Gibson AI (2002) Mitigation options for greenhouse gas emissions from agriculture. University of London, London

Haberl H, Sprinz D, Bonazountas M, Cocco P, Desaubies Y, Henze M, Lange E (2012) Correcting a fundamental error in greenhouse gas accounting related to bioenergy. Energy Policy 45:18–23

Herrero M, Thornton PK, Kruska R, Reid RS (2008) Systems dynamics and the spatial distribution of methane emissions from African domestic ruminants to 2030. Agric Ecosyst Environ 126(1–2):122–137

Hoekman SK, Broch A (2018) Environmental implications of higher ethanol production and use in the US: a literature review. Part II–biodiversity, land-use change, GHG emissions, and sustainability. Renew Sust Energ Rev 81:3159–3177

Huang R, Wang Y, Liu J, Li J, Xu G, Luo M, Gao M (2019) Variation in N2O emission and N2O related microbial functional genes in straw-and biochar-amended and non-amended soils. Appl Soil Ecol 137:57–68

Inglesi-Lotz R, Dogan E (2018) The role of renewable versus non-renewable energy to the level of CO2 emissions a panel analysis of sub-Saharan Africa's Big 10 electricity generators. Renew Energy 123:36–43

Joint FAO (2018) Rice production guidelines: best farm management practices and the role of isotopic techniques (No. IAEA-TECDOC–1847). Joint FAO/IAEA Division of Nuclear Techniques in Food and Agriculture

Kumar S, Meena RS, Datta R, Verma SK, Yadav GS, Pradhan G, . . . Mashuk HA (2020) Legumes for carbon and nitrogen cycling: an organic approach. In: Carbon and nitrogen cycling in soil. Springer, Singapore, pp 337–375

Leng LY, Ahmed OH, Jalloh MB (2019) A brief review on climate change and tropical peatlands. Geosci Front 10(2):373–380

Lovett DK, Shalloo L, Horan B, Dillon P, O'Mara FP (2006) Effect of Holstein–Friesian strain and feeding system on greenhouse gas emissions from pastoral dairy production systems. In: International Congress Series, vol 1293. Elsevier, pp 335–338

Malaka SF (2017) Estimation of greenhouse gas emissions from agriculture in the eastern Free State, South Africa (Doctoral dissertation, University of the Free State)

Martinez-Fernandez G, Duval S, Kindermann M, Schirra HJ, Denman SE, McSweeney CS (2018) 3-NOP vs. halogenated compound: methane production, ruminal fermentation and microbial community response in forage-fed cattle. Front Microbiol 9:1582

Mehra P, Baker J, Sojka RE, Bolan N, Desbiolles J, Kirkham MB, Gupta R (2018) A review of tillage practices and their potential to impact the soil carbon dynamics. In: Advances in agronomy, vol 150. Academic Press, pp 185–230

Mottet A, Henderson B, Opio C, Falcucci A, Tempio G, Silvestri S, Gerber PJ (2017) Climate change mitigation and productivity gains in livestock supply chains: insights from regional case studies. Reg Environ Chang 17(1):129–141

Müller A, Bautzen L, Meier M, Gattinger A, Gall E, Chatzinikolaou E, Ullmann L (2016) Organic farming, climate change mitigation and beyond. Reducing the environmental impacts of EU agriculture. IFOAM EU, Brussels

Mwanake RM, Gettel GM, Aho KS, Namwaya DW, Masese FO, Butterbach-Bahl K, Raymond PA (2019) Land use, not stream order, controls N2O concentration and flux in the upper Mara River basin, Kenya. J Geophys Res Biogeo 124(11):3491–3506

Novak SM, Fiorelli JL (2010) Greenhouse gases and ammonia emissions from organic mixed crop-dairy systems: a critical review of mitigation options. Agron Sustain Dev 30(2):215–236

Nyamoga GZ, Solberg B (2019) A review of studies related to charcoal production, consumption, and greenhouse gas emissions in Tanzania. In: Agriculture and ecosystem resilience in Sub Saharan Africa. Springer, Cham, pp 357–399

Oberthür S, Hermwille L, Khandekar G, Obergassel W, Rayner T, Wyns T, and Melkie M (2019) Key concepts, core challenges and governance functions of international climate governance: deliverable 4.1; COP21-results and implications for pathways and policies for low emissions European societies

Owen JJ, Silver WL (2015) Greenhouse gas emissions from dairy manure management: a review of field-based studies. Glob Chang Biol 21(2):550–565

Paustian K, Larson E, Kent J, Marx E, Swan A (2019) Soil C sequestration as a biological negative emission strategy. Front Clim 1(8):1–11

Popp A, Lotze-Campen H, Bodirsky B (2010) Food consumption, diet shifts and associated non-CO2 greenhouse gases from agricultural production. Glob Environ Chang 20(3):451–462

Ronaghi M, Saghaian S, Reed M, Mohammadi H (2018) The impact of the agricultural sector in developing countries that produce natural gas on greenhouse gas emissions. Int J Food Agric Econ (IJFAEC) 6(1128–2019-555):53–69

Saha S, Chatterjee D, Swain CK, Nayak AK (2018) Methane emission from wetland rice agriculture-biogeochemistry and environmental controls in projected changing environment. In: Advances in crop environment interaction. Springer, Singapore, pp 51–85

Sanz-Cobena A, Lassaletta L, Aguilera E, Del Prado A, Garnier J, Billen G, Plaza-Bonilla D (2017) Strategies for greenhouse gas emissions mitigation in Mediterranean agriculture: a review. Agric Ecosyst Environ 238:5–24

Šimek M, Cooper JE (2002) The influence of soil pH on nitrification: progress towards the understanding of this interaction over the last 50 years. Eur J Soil Sci 53:345–354

Slaets J, Mayr L, Heiling M, Zaman M, Resch C, Weltin G, Dercon G (2016) Protocol development for continuous nitrogen-15 measurement of N_2O and its isotopomers for real-time greenhouse gas tracing

Smith P, Martino D, Cai Z, Gwary D, Janzen H, Kumar P, Scholes B (2008) Greenhouse gas mitigation in agriculture. Philos Trans R Soc B Biol Sci 363(1492):789–813

Stanton CY, Mach KJ, Turner PA, Lalonde SJ, Sanchez DL, Field CB (2018) Managing cropland and rangeland for climate mitigation: an expert elicitation on soil carbon in California. Climatic Change 147(3–4):633–646

Swamy M, Bhattacharya S (2006) Budgeting anthropogenic greenhouse gas emission from Indian livestock using country-specific emission coefficients. Curr Sci:1340–1353

Tongwane MI, Moeletsi ME (2018) A review of greenhouse gas emissions from the agriculture sector in Africa. Agric Syst 166:124–134

Wall E, Simm G, Moran D (2010) Developing breeding schemes to assist mitigation of greenhouse gas emissions. Animal 4(3):366–376

Wang Y, Dong H, Zhu Z, Gerber PJ, Xin H, Smith P, Chadwick D (2017) Mitigating greenhouse gas and ammonia emissions from swine manure management: a system analysis. Environ Sci Technol 51(8):4503–4511

Winnie N, Gweyi-Onyango J (2018) Biofertilizers as an alternative to inorganic fertilizers in sub-Saharan African: is the adoption the missing link? Eliexir Agric 114:49564–49551

Yang Y, Tilman D, Furey G, Lehman C (2019) Soil carbon sequestration accelerated by the restoration of grassland biodiversity. Nat Commun 10(1):1–7

Zerhusen B, Maze M, Effenberger M (2019) Hands-on model for assessing the green-house gas emissions and the fossil energy consumption of agricultural biogas plants. Landtechnik 74(5):118–134

Zhu XC, Di DR, Ma MG, Shi WY (2019) Stable isotopes in greenhouse gases from soil: a review of theory and application. Atmosphere 10(7):377

4

Precipitation Variability in West Africa in the Context of Global Warming and Adaptation Recommendations

Gandome Mayeul L. D. Quenum, Nana A. B. Klutse, Eric A. Alamou, Emmanuel A. Lawin and Philip G. Oguntunde

Contents

G. M. L. D. Quenum (✉)
West African Science Service Centre for Climate Change and Adapted Land Use (WASCAL) Graduate Research Program in West African Climate System (GRP-WACS), Federal University of Technology, Akure (FUTA), Akure, Nigeria

Laboratory of Applied Hydrology (LHA), National Institute of Water (NIW), Cotonou, Bénin
e-mail: malgdaq2000@yahoo.fr

N. A. B. Klutse
Ghana Space Science and Technology Institute, Atomic Energy Commission, Accra, Ghana

E. A. Alamou · E. A. Lawin
Laboratory of Applied Hydrology (LHA), National Institute of Water (NIW), Cotonou, Bénin

P. G. Oguntunde
Department of Agricultural and Environmental Engineering, Federal University of Technology, Akure, Nigeria

Abstract

It is commonly accepted that the Earth's climate is changing and will continue to change in the future. Rising temperatures are one of the direct indicators of global climate change. To investigate how the rising global temperature will affect the spatial pattern of rainfall in West Africa, the precipitation and potential evapotranspiration variables from ten Global Climate Models (GCMs) under the RCP8.5 scenario were driven by the Rossby Centre regional atmospheric model (RCA4) from the COordinated Regional Climate Downscaling EXperiment (CORDEX) and analyzed at four specific global warming levels (GWLs) (i.e., 1.5 °C, 2.0 °C, 2.5 °C, and 3.0 °C) above the preindustrial level. This study utilized three indices, the precipitation concentration index (PCI), the precipitation concentration degree (PCD), and the precipitation concentration period (PCP) over West Africa to explore the spatiotemporal variations in the characteristics of precipitation concentrations. Besides, the analysis of the effect of the specified GWLs on the Consecutive Dry Days (CDD), Consecutive Wet Days (CWD), and frequency of the intense rainfall events allowed to a better understanding of the spatial and temporal patterns of extreme precipitation in West Africa. Results reveal that, for the projections simulations and at each GWL, the rainfall onset starts one month earlier in the Gulf of Guinea in response to the control period. To encourage adaptation to the various changes in climate in general, and particularly in respect of rainfall, this study proposes several adaptation methods that can be implemented at the local (country) level, as well as some mitigation and adaptation strategies at the regional (West African) level.

Keywords

Precipitation concentration index (PCI) · Precipitation concentration degree (PCD) · Precipitation concentration period (PCP) · Adaptation strategy · Global warming

Introduction

Precipitation variability and concentration are two important climate factors with an impact on society, agriculture, and the environment, as increased precipitation variability can reduce agricultural yields (Rowhani et al. 2011). Also, precipitation variability is strongly connected to extreme wet and dry events, floods, and droughts, which pose threats to the environment and to society, and can also have devastating consequences on ecosystems, food supplies, and economies at the local, regional, and global scale (Easterling et al. 2000). In its 5th report, the Intergovernmental Panel on Climate Change (IPCC) stresses that there is an increase in the number of extreme weather events for the twenty first century due to climate change (IPCC 2014). As temperatures rise, the amount of water vapor in the atmosphere also increases and the spatiotemporal distributions of precipitation change, resulting in significant differences in precipitation across the world (Chou and Lan 2011;

Gao et al. 2014). Obtaining real-time information and facilitating earlier predictions by decision makers can be an efficient tool to adapt and to mitigate the impacts of climate events. However, this is an especially challenging task over data-sparse regions as West Africa, where unreliable monitoring networks and generally low institutional capacity limits the spread of timely information (Sheffield et al. 2014).

To assess the spatial and temporal variability of precipitation, various methods were adopted and documented. Precipitation concentration plays an important role, firstly, with regard to the annual total precipitation amount days and, secondly, with regard to the time and degree of concentration of the yearly total precipitation within a year. This variable has the potential to cause floods and drought events, both of which are expected to put considerable pressure on water resources (Zhang and Qian 2003; Zhang et al. 2009). Several studies have been interested in the importance of precipitation concentration indices. The present study adopted three of them: (i) the precipitation concentration index (PCI), which was developed by Oliver (1980); and years after was modified by De Luis et al. (2011). It has been used by Shi et al. (2015) as an indicator of the rainfall concentration for annual and seasonal scales, and Ezenwaji et al. (2017) to investigate the implications of the concentration and variability of rainfall on flooding over Awka Urban Area (Nigeria). (ii) The precipitation concentration degree (PCD) and (iii) the precipitation concentration period (PCP), which were redefined by Zhang and Qian (2003) in a study of droughts and floods in the Yangtze River valley (China). The basic concept behind the PCD and PCP is that the monthly total precipitation is a vector quantity with both magnitude and direction (Li et al. 2011; Wang et al. 2013).

In order to investigate and understand future climates, scientists refer to climate scenarios to provide a plausible explanation of how the future may evolve with respect to a number of variables, including socioeconomic change, technological change, energy and land-use, and the emissions of greenhouse gasses (GHGs) and air pollutants (Van Vuuren et al. 2011). Many factors are taken into account in order to predict how future global warming will contribute to climate change. A key variable is future GHG emissions. Some studies over West Africa (e.g., Egbebiyi 2016; Haensler et al. 2013; Sylla et al. 2012) have shown that projected increases in global GHG concentrations will lead to an increase in the frequency and intensity of extreme rainfall events (Klutse et al. 2018). So, for the meeting by the Intergovernmental Panel on Climate Change (IPCC) in 2007, four RCP radiative forcing levels were retained (Moss et al. 2008). The Representative Concentration Pathway (RCP) is defined as the trajectories relating to GHG concentrations, and was projected by the IPCC in 2014. The four Representative Concentration Pathways (RCPs), namely, RCP2.6, RCP4.5, RCP6, and RCP8.5, are labeled after a possible range of radiative forcing values up to the year 2100 (2.6, 4.5, 6.0, and 8.5 W/m^2, respectively).

In 2010, the sixteenth session of the Conference of the Parties (COP16) adopted the Cancun Agreement for keeping the global temperature rise below 2.0 °C (UNFCCC 2010). A few years later, in 2015, at Paris, the COP21 proposed an ambitious target, which is to keep the increase in global average temperature to well below 2.0 °C above preindustrial levels, while pursuing efforts to limit the temperature increase to 1.5 °C. The rate of climate change is an important issue for mitigation and/or adaptation policies (Watkiss et al. 2015). Many studies have recently examined regional risks for the whole of Africa, as well some of its regions, such as Southern Africa and West Africa, at various global warming levels (GWLs). Recently, Déqué et al. (2017) suggested a classification of global warming for all the RCPs and for GWLs from 1.5 °C to 4 °C. Based on Déqué et al.'s (2017) suggestion, several studies (Abiodun et al. 2019; Klutse et al. 2018; Kumi and Abiodun 2018; Maúre et al. 2018; Nikulin et al. 2018) were done to assess the impact of different GWLs on various climate variables.

Mitigation and adaptation are two fundamental societal response options to deal with the problem of climate change (IPCC 2001; Füssel 2007; Locatelli 2011). The community of climate change scientists has focused mainly on mitigation rather than adaptation; for a long time, the latter has been ignored in the debate on climate change (Füssel 2007; De Perthuis et al. 2010). Mitigation policies have already been adopted at the international level, since it is clear that the global climate is already changing and will continue to change in the future due to anthropogenic GHG and aerosol emissions; it is thus important to make adjustments all the more urgently, especially at a local level (Füssel 2007; De Perthuis et al. 2010). In addition, though, there is a need to focus the debate on adaptation measures to climate change in order to prepare communities, especially the most vulnerable ones, to cope with its impacts (Kpadonou et al. 2012).

Identification of a suitable methodology to assess the spatiotemporal precipitation distribution is vital for planning effective adaptation and mitigation strategies to overcome potential drought and flood situations. Selecting the optimal methodology or tools may be useful for planning disaster management and preparedness to respond to droughts and floods, and to mitigate their effects on the activities in different sectors of the economy. Therefore, this study aims to use the analyses of PCI, PCD, PCP, consecutive dry days (CDD), and consecutive wet days (CWD) to explain the spatial and temporal variability of precipitation in West Africa in the context of GWLs 1.5 °C, 2.0 °C, 2.5 °C, and 3.0 °C.

Description of the Study Areas and Methodology

This section describes the study area and the data used (section "Study Area and Data Sources"), and the methodology that was applied (section "Methodology") to arrive at the results.

Study Area and Data Sources

The study area lies in West Africa, which is located between latitudes 0°N and 20°N and longitudes 20°W and 20°E (Fig. 1). This region is bordered in the South by the Gulf of Guinea, in the north by Mauritania, Mali, and Niger; the Cameroon highlands form the eastern boundary, while the Atlantic Ocean forms the western limit. The annual mean temperature is about 18 °C, but the monthly mean can be more than 40 °C over the southern part of the Sahara. Rainfall patterns over this region are mostly affected by ocean currents and local features, such as topography.

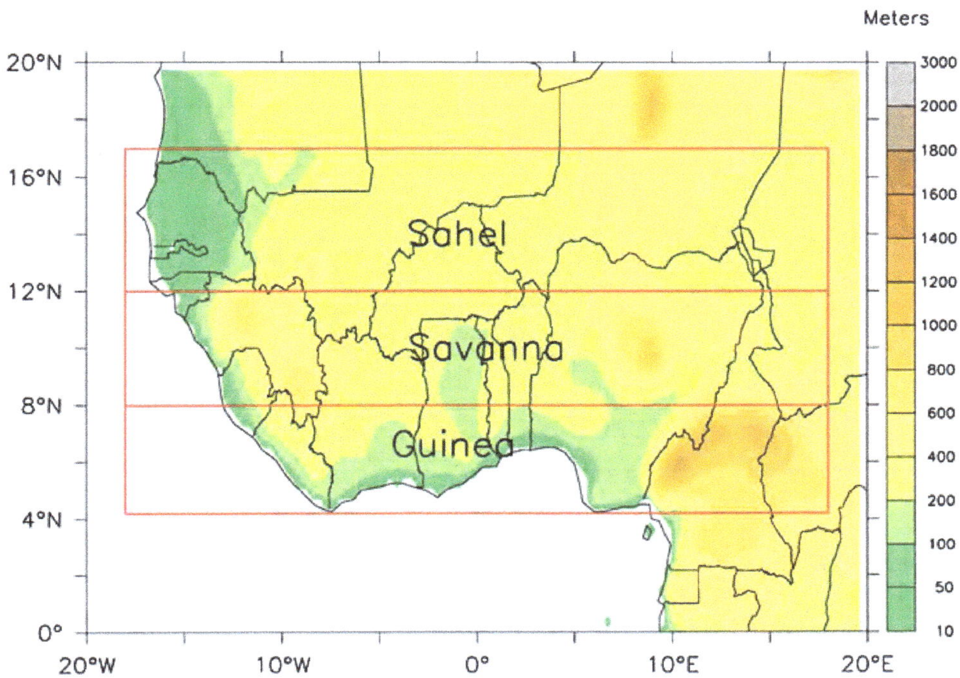

Fig. 1 Study domain showing the West African topography and the area of focus, which comprises the Gulf of Guinea (Guinea), the Savanna, and the Sahel zones (Diasso and Abiodun 2015)

In terms of climatic zones, West Africa can be divided into three different regions. The first region covers the Sahel and is characterized as a semiarid zone ranging from western Senegal to eastern Sudan, between 12°N and 20°N. The second region is the Sudano-Sahelian zone, while the last and the third comprises the Guinea coast, which is characterized by a bimodal mode driven by the Intertropical discontinuity (ITD).

Precipitation and potential evapotranspiration variables from CORDEX are used in this study. Data from CORDEX are driven by RCA4 at daily and monthly timesteps. The simulated dataset used was collected from the coordinated regional climate downscaling experiment (CORDEX) (Nikulin et al. 2012), and has a horizontal resolution of 0.44° × 0.44°. The period of 1971–2000 was extracted from the simulated dataset as the historical period, while the period of 2006–2100 was used for projections. This study focuses specifically on the GWLs of 1.5 °C, 2.0 °C, 2.5 °C, and 3.0 °C above the preindustrial levels and under the RCP8.5 scenario. The 30-year projection according to each GWL can be found in Table 1.

Methodology

The PCI is calculated in the section "Calculation of Precipitation Concentration Index (PCI)" to classify how the rainfall is spatially distributed, and the PCD and PCP in the section "Computation of Precipitation Concentration Degree (PCD) and Precipitation

Table 1 Recapitulative of the selected years for the ten GCMs drove by RCA4 model according to each global warming level. (Adapted by Quenum et al. 2019 from Déqué et al. 2017)

	RCP85				
RCM	GCM	1.5 °C	2.0 °C	2.5 °C	3.0 °C
RCA4	CanESM2	1999–2028	2012–2041	2024–2053	2034–2063
	CNRM-CM5	2015–2044	2029–2058	2041–2070	2052–2081
	CSIRO-Mk3	2018–2047	2030–2059	2040–2069	2050–2079
	EC-EARTH-r12	2005–2034	2021–2050	2034–2063	2047–2076
	GFDL-ESM2M	2020–2049	2037–2066	2052–2081	2066–2095
	HadGEM2-ES	2010–2039	2023–2052	2033–2062	2042–2071
	IPSL-CM5A-MR	2002–2031	2016–2045	2027–2056	2036–2065
	MIROC5	2019–2048	2034–2063	2047–2076	2058–2087
	MPI-ESM-LR	2004–2033	2021–2050	2034–2063	2059–2088
	NorESM1-M	2019–2048	2034–2063	2047–2076	2059–2088

Concentration Period (PCP)" were designed to determine, respectively, the period (months) and the degree of the concentration of the precipitation during the year.

Calculation of Precipitation Concentration Index (PCI)

The PCI, which was developed by Oliver (1980), modified by De Luis et al. (2011) and also used by Shi et al. (2015), was used as an indicator of rainfall concentration for annual and seasonal scales (wet and dry seasons). In our study, the PCI was tested to identify future trends in respect of the spatial distribution of rainfall. According to Oliver (1980), PCI values of less than 10 represent a uniform precipitation distribution (i.e., a low precipitation concentration), values between 11 and 15 denote a moderate precipitation concentration, values from 16 to 20 denote an irregular distribution, and values above 20 represent a strong irregularity of precipitation distribution (i.e., a high precipitation concentration). The following equations were used on each grid point to calculate the PCIs:

$$\mathrm{PCI}_{\mathrm{annual}} = \frac{\sum_{i=1}^{12} P_i^2}{\left(\sum_{i=1}^{12} P_i\right)^2} \times 100 \tag{1}$$

$$\mathrm{PCI}_{\mathrm{wet}} = \frac{\sum_{i=1}^{nw} P_i^2}{\left(\sum_{i=1}^{nw} P_i\right)^2} \times \frac{100 * nw}{12} \tag{2}$$

$$\mathrm{PCI}_{\mathrm{dry}} = \frac{\sum_{i=1}^{nd} P_i^2}{\left(\sum_{i=1}^{nd} P_i\right)^2} \times \frac{100 * nd}{12} \tag{3}$$

Equation 1 was used for annual PCI, while Equations 2 and 3 were utilized for seasonal scales (rainy and dry seasons, respectively); *nw* and *nd* represent, respectively, the number of rainy and dry season months, and $P =$ precipitation of the ith month. In order to investigate changes in the PCI, a 30-year period was considered both for historical and future periods. Table 1 shows the projection periods used for each GWL; the historical 30-year period is 1971–2000.

Computation of Precipitation Concentration Degree (PCD) and Precipitation Concentration Period (PCP)

The PCD and PCP were proposed by Zhang and Qian (2003) to measure the distribution of rainfall and the peak of its concentration date. The basic principle is based on the vector of daily or monthly total precipitation. The assumption can be made that at a time scale (daily, 5-day, weekly, decade or monthly) total precipitation is a vector quantity with both magnitude and direction and can be illustrated as a circle (360°). According to Li et al. (2011) and Zhang and Qian (2003), the indices were calculated as follows:

$$\theta_j = \left(360^\circ * \frac{j}{n}\right) \tag{4}$$

$$R_i = \sum r_{ij} \tag{5}$$

$$R_{xi} = \sum_{j=1}^{N} r_{ij} * \sin\theta_j \tag{6}$$

$$R_{yi} = \sum_{j=1}^{N} r_{ij} * \cos\theta_j \tag{7}$$

$$\mathrm{PCD}_i = \frac{\sqrt{{R_{xi}}^2 + {R_{yi}}^2}}{R_i} \tag{8}$$

where i is the year (e.g., for the historical period $i = 1971, 1972, \ldots, 2000$), j represents the time scale (**daily**, 5-day, weekly, decade, or **monthly**) of that year, R_i is the amount of rainfall of a year, r_{ij} is the precipitation of the jth time scale in the ith year, n is the number of time scales per year (e.g., daily: for a non-leap year, $n = 365$, while in a leap year, $n = 366$)

$$\alpha_i = \tan^{-1}\left(\frac{R_{xi}}{R_{yi}}\right) \tag{9}$$

$$D_i = \begin{cases} \alpha_i \ (R_{yi} > 0, R_{xi} \geq 0) \\ \alpha_i \ (R_{yi} > 0, R_{xi} < 0) \\ \alpha_i \ (R_{xi} < 0) \end{cases} \tag{10}$$

$$\mathrm{PCP}_i = D_i * \left(\frac{n}{360^\circ}\right) \tag{11}$$

The next section discusses the results obtained with regard to the three indices above mentioned.

Results

This section presents the results obtained from the study. Based on the variability between the historical period (1971–2000, also referred to as the control period, CTL) and the various GWLs periods the section "Annual Precipitation Concentration" shows the annual precipitation concentration patterns, while the section "Seasonal Precipitation Concentration" focuses on the seasonal PCI variabilities. The section "Evaluation of the Models' Robustness" investigates the robustness of the models to assess the PCI over West Africa. The period and the degree of the concentration of the rainfall are evaluated in the section "Variability of PCD and PCP", while the section "Daily Precipitation Variability" shows the daily consecutive precipitation variability. Finally, the adaptation processes proposed are explained in section "Adaptation Strategies."

Variability of PCI

The variability of the PCI is investigated in sections "Annual Precipitation Concentration" and "Seasonal Precipitation Concentration," respectively, for the annual and seasonal time scales.

Annual Precipitation Concentration

The variability of the PCI computed across West Africa at an annual scale is between 12 and above 20 (Fig. 2a–e). According to a classification established by Oliver (1980), this high variability of the PCI values over West Africa illustrates the existence of a seasonal rainfall regime. Lower values (between 12 and 13) are noticed for the simulation of the historical period over the Gulf of Guinea, which means that this region of the study domain has a moderate precipitation concentration when interested in the distribution at the annual scale. The seasonality is more pronounced in the Savanna with PCI range between 17 and 18 (according to Oliver's (1980), this explains how the precipitation is irregularly distributed both spatially and temporally). Lastly, the Sahel region recorded a high precipitation concentration (PCI>20), which means that the precipitation is strongly irregularly distributed. For the specified GWLs, it is illustrated in Fig. 2a–e that for the Gulf of Guinea and the Savanna, the precipitation concentration is irregularly distributed except for some countries like Côte d'Ivoire and Liberia which have a low PCI.

Seasonal Precipitation Concentration

There are two major seasons over the study area. For the purposes of this analysis, the rainy season is assumed to last from early May to the end of September (MJJAS).

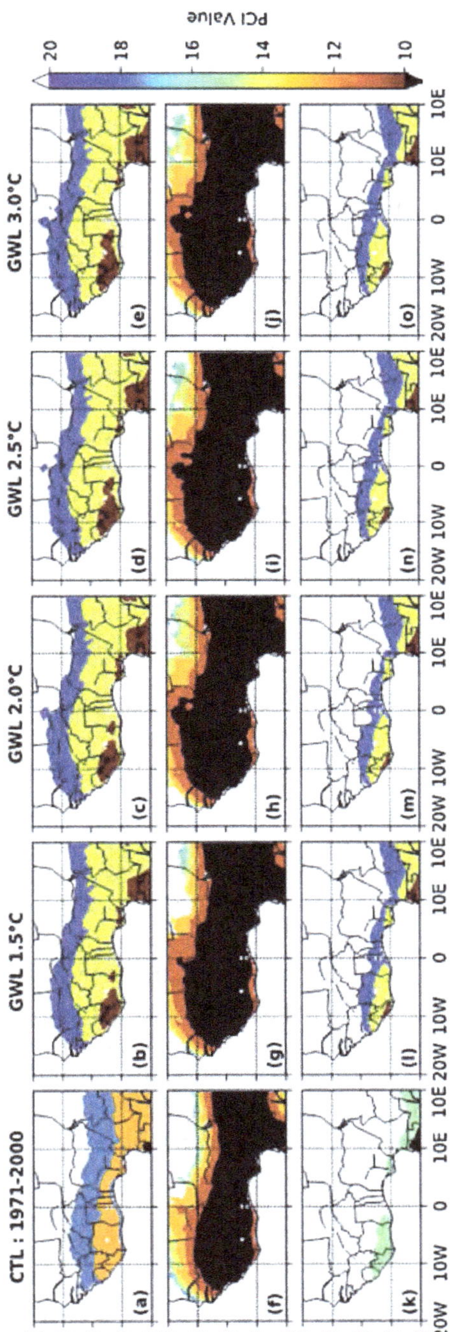

Fig. 2 Variability of Precipitation Concentration Index (PCI) at annual and seasonal scales for the historical period (CTL: 1971–2000) and for projections of GWLs 1.5 °C, 2.0 °C, 2.5 °C, and 3.0 °C

The PCI computed for the seasonal scale over West Africa indicates complex spatial patterns of precipitation distribution. Therefore, Fig. 2f–j shows that precipitation is uniformly distributed (i.e., almost the same amount of precipitation occurs in each month) over the Gulf of Guinea and the Savanna. For the studied GWLs, when compared with the historical period, it can be noticed that the uniform precipitation distribution extends a little toward the Sahel. So, the precipitation concentration is irregularly distributed (i.e., $PCI \in [16, 20]$ according to Oliver 1980) over the northern region of West Africa during the rainy season.

Figure 2k–o presents the PCI for the dry season. It illustrates an irregular precipitation concentration distribution over the Gulf of Guinea. This is obvious, because there is almost no rain during this period of the year, and the irregular distribution noticed might be due to some isolate rains recorded after some particular meteorological phenomena. The precipitation in the Savanna and Sahel is strongly irregularly distributed (i.e., $PCI > 20$ during the rainy period, according to Oliver 1980), which means that the total of significant precipitation occurs within a short period (1–2 months).

The analysis of simulations presented in Fig. 2 (with regard to annual and seasonal evaluation) establishes that in West Africa, the precipitation is uniformly well distributed during the period May–September (MJJAS) in the Gulf of Guinea and the Savanna. A northward gradient is well noticed because the highest values of the PCI are located in the Sahel, whereas the lowest are identified around the Gulf of Guinea. Despite the global warming effect for all levels, the precipitation concentration does not change over the Gulf of Guinea and the Savanna; on the contrary, it extends toward the northern region of the study domain.

Evaluation of the Models' Robustness

Figure 3, which presents the differences between the projected PCI in respect of the historical period, shows that the level of variability is similar from one GWL to another. The annual and seasonal concentrations reduce gradually from the Sahel to the Gulf of Guinea, and confirm the variability illustrated by Fig. 2, which shows the regression of irregular and strong irregular precipitation concentrations. Figure 3 also illustrates the robustness of the simulations. At least 80% of models (indicated here with vertical green strips) demonstrate that the precipitation concentration over the eastern part of the study area has changed. This change, which increases according to the GWLs, is also shown over several countries, such as Niger during the rainy season. At least 80% of the models demonstrated that the change is significant (as indicated by the horizontal blue strips), with a confidence level of 95%. Here too, Niger and Nigeria are projected to experience significant changes, which will increase with the GWLs. The red cross (+) is observed in the area where at least 80% of the simulations agree with regard to the change, and where these changes have a 95% confidence level. Therefore, during the rainy season and under GWL3.0, countries such as Ghana, Togo, and Burkina Faso present a more uniform precipitation distribution, in contrast to variabilities for the historical period and the projections (Fig. 2).

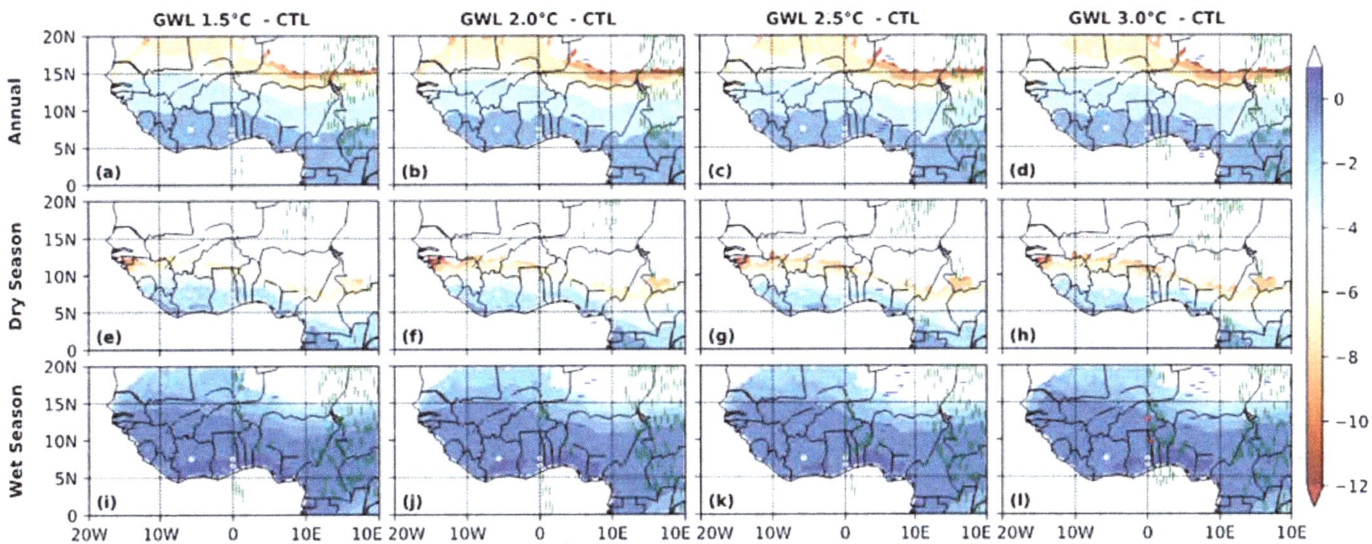

Fig. 3 Evaluation over West Africa of the difference between the changes at 1.5 °C, 2.0 °C, 2.5 °C, and 3.0 °C GWLs of PCI using the CORDEX Africa ensemble. The vertical green strip (|) indicates where at least 80% of the models agree on the sign of the changes, while horizontal blue strip (–) indicates where at least 80% of the simulations agree that the projected change is statistically significant with 95% as confidence level. The red cross (+) indicates where both conditions are satisfied

Variability of PCD and PCP

Figure 4 presents the PCP and PCD. Figure 4a–e shows that the variability over the entire West Africa of PCPs both for historical and projected periods is a range between 7 ± 2. This indicates that the yearly mean precipitation concentration over West Africa occurs between June and September. The analysis corroborates with the current knowledge about the period of the West African precipitation producing system, which is led by the West African Monsoon (WAM). The high values during the control period are observed in the north-western of West Africa, while regarding the projected period, the highest values are recorded over the Sahel. The finding indicates that the wet season starts earlier in the southern regions, followed by the Savanna before reaching the Sahel in the North. The variability of the yearly mean PCDs (shown in Fig. 4f–j) is between 0.17 and 0.90, indicating the high spatial variability of the precipitation concentration in the study domain. When looking at the historical period simulations (Fig. 4f), it can be observed that the computed PCDs increase northwardly. Lower values (0.17–0.60) are located around the Guinea coast, and the highest (>0.80) in the Sahel. This denotes the existence of a gradient across the Gulf of Guinea and the Sahel. This gradient informs that precipitation during the rainy season is more concentrated in a few months across the Sahel compared to the Gulf of Guinea. For the projected simulations, the same gradient Gulf-of-Guinea-Sahel is observed, although here the magnitude of the PCD is less in response to the control period. The lower values lie between 0.17 and 0.50, and the higher range between 0.5 and 0.6. During the projected period, there is globally a decrease of precipitation concentration compared to the historical period, leading both the Sahel and the Savanna having the same precipitation distribution for all GWLs studied. Furthermore, the projections indicate a shift in time for the starting of the rainy season. Therefore, comparing the projected and the present period, it is noticed that the wet season will state earlier. The highest concentration of precipitation will occur from May to July for the Gulf of Guinea and the Savanna, and in August for the Sahel.

Daily Precipitation Variability

The consecutive wet days (CWD) and consecutive dry days (CDD) were calculated over the study domain to evaluate the daily variability of the precipitation distribution. CWD and CDD also indicate extremes in rainfall. CDD is furthermore a useful indicator for studying short-term droughts (Frich et al. 2002) and drought tendencies (Orlowsky and Seneviratne 2012), as it could indicate enhanced dryness and high risk for seasonal droughts (Klutse et al. 2018). Changes in CDD and CWD can lead to uneven temporal distributions of rainfall, which could have a significant consequence for agricultural practices (Barron et al. 2003; FAO et al. 2015; Wiebe et al. 2017). The CDD was calculated both at annual (cdd) and seasonal scales (in this study, May–September: MScdd), in order to evaluate both dry and wet spells within the rainfall season; knowing this is very important for agricultural practices in the region (Klutse

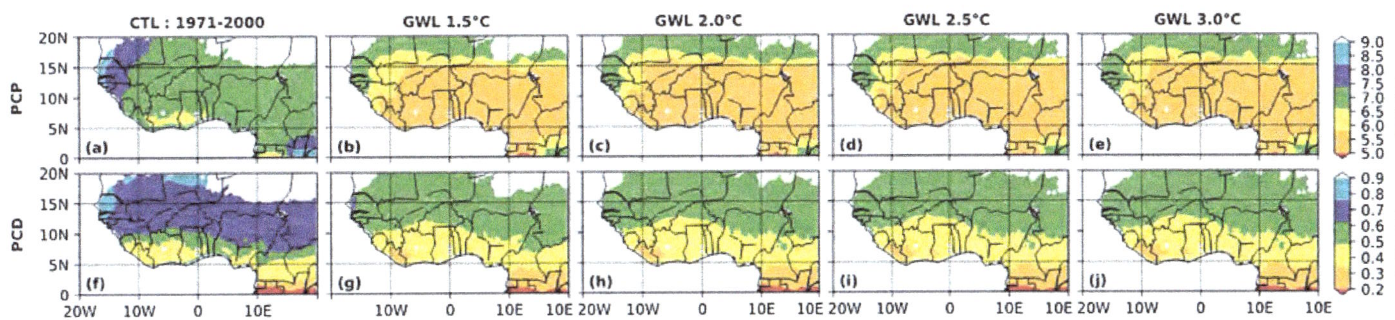

Fig. 4 Spatial distribution of yearly mean Precipitation Concentration Degree (PCD) and Precipitation Concentration Period (PCP) for the historical period (CTL: 1971–2000) and at projections of GWLs 1.5 °C, 2.0 °C, 2.5 °C, and 3.0 °C

et al. 2018). Figure 5 shows the variability between the projection of different GWLs and the selected historical period. High values of CDD are noticed in the north of West Africa, and higher CWD values around the coast of Gulf of Guinea. The comparison between the results of (Figs. 5a–d, i–l) shows both for annual and rainy season periods a decrease of CDD about 10 ± 5 days over the north-eastern of West Africa. Around the northern area of the study domain, an important variability of dry days is noticed within the rainy season (e.g., a decrease in CDD over the north-eastern part and an increase in CDD over the north-western part), which demonstrates that the north-eastern of West Africa is wetter under GWLs, while the north-western region is drier. Over the Gulf of Guinea, for both annual and rainy season scales a slight variability of CDD for all GWLs is noticed. The results of the GWLs 1.5 °C, 2.0 °C, and 2.5 °C indicate a pattern of CDD with same spatial variability, and for the GWL 3.0 °C (on Fig. 5i), a significant increase in the annual CDD is observed. The CDD is projected to increase for 4–5 days over the Gulf of Guinea; in Mauritania and Senegal, the increase is projected to be 10 ± 2 days as showed by Quenum et al. (2019). Some Sahel countries like Niger and Chad (which are characterized by a dry north-easterly flow crossing the Sahara desert) are projected to record a dwindling of CDD between 10 and 14 days. This finding agrees at GWLs 1.5 °C and 2.0 °C, with Klutse et al. (2018), who showed a reduction for GWLs 1.5 °C and 2.0 °C, in terms of the number of CDD in the wet season over our study domain, and also the results of Sultan and Gaetani (2016), who concluded a decrease in the number of CDD over central Africa.

In contrast to the CDD analysis over West Africa, the CWD did not show important variability. Its variation is very slight and ranges between 0 ± 3 days. Nonetheless, some high variations could be observed at several specific points. On the projected simulations illustrated in Fig. 5e–h), the CWD decreases about 10 ± 2 over the South Benin and Nigeria. A slight augment in CWD of up to 2 days is likely to be noticed in the Sahel.

In order to investigate the spatial variability of extreme rainfall events, which plays an important role in the availability of water resources and agriculture, etc., the frequency of intense rainfall events (RxD10 mm: $R \geq 10$ mm/day), very intense rainfall events (RxD20 mm: $R \geq 20$ mm/day), and heavy rainfall events (RxD25 mm: $R \geq 25$ mm/day) were calculated; they are displayed in Fig. 6. These variables indicate whether there were changes in the amount of precipitation received over consecutive 5 days with the highest precipitation. Figure 6a–d illustrate that, compared to the control period, each GWL detects an increasing RxD10 mm over the orographic regions and the ocean boundary (Gulf of Guinea). There is a very slight increase in the number of RxD10 mm over the Savanna and Sahel zones. In general, the results clearly show that as the GWL increases, the more the projected RxD10 mm increases too (e.g., for GWL 1.5 °C, the increase is about 7 ± 2 over the Gulf of Guinea and 1 ± 1 for the Savanna and Sahel, while for GWL 3.0 °C, the increase is about 9 ± 2 over the Gulf of Guinea and 3 ± 1 for the Savanna and Sahel). In the case of RxD20 mm and RxD25 mm, the general increase in response to increasing GWLs is noticed too. Only the coastal countries record significant increases in RxD20 mm and RxD25 mm, which could be due to the south-westerly moist flow from the Gulf of Guinea inland.

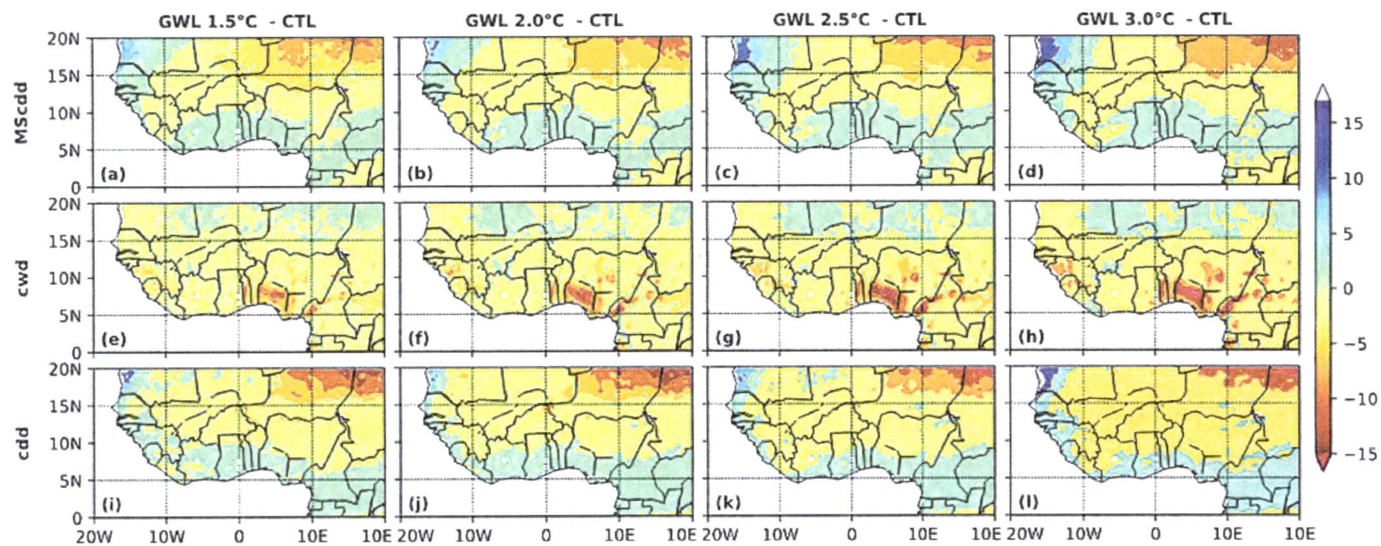

Fig. 5 Spatial distribution of the change in Consecutive Dry Day (CDD) during the rainy season (MScdd), as well as the annual consecutive dry days (cdd) and the consecutive wet days (cwd)

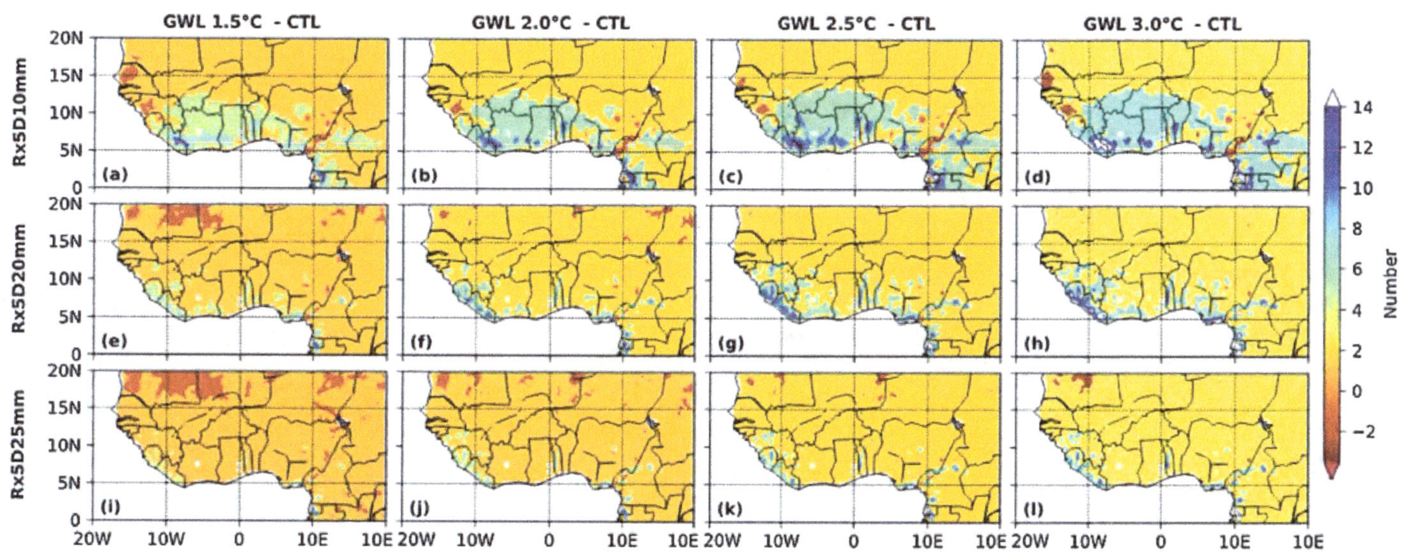

Fig. 6 Spatial distribution of the change in frequency of intense rainfall events (RxD10 mm), very intense rainfall events (RxD20 mm), and heavy rainfall events (RxD25 mm)

Adaptation Strategies

Since the spectacular drought events of the 1970s, it has become clear that the high variability in precipitation constitutes one of the major challenges faced by the West African region. Agriculture is one of the major economic activities of West Africa, and thus significant changes in rainfall due to climate change will negatively affect the entire region. These concerns have generated ongoing scientific, social, and political debate. Moreover, some parts of West Africa (mostly along the Guinean Coast) have recorded recurrent flood events since 2000. Thus, both climate variability and increasing trends in droughts and floods and other severe weather events pose a challenge for the primarily rain-fed agriculture systems in West Africa (Sultan and Gaetani 2016). Therefore, any adaptations must enable inhabitants to cope successfully with short-term climate variability as well as to reduce the long-term negative impacts of climate change (Lobell 2014; Saba et al. 2013). Households and communities must become accustomed to and able to respond creatively and effectively to disruptions of their livelihoods. Indeed, in order to be successful, adaptations must be anchored in all processes affecting life. Some of the possible adaptation strategies, especially relating to floods, droughts, and food crops, are illustrated in this study.

According to the results above, at increasing GWLs, precipitation over the Gulf of Guinea and the Savanna will shift and start earlier (in May) and that the highest precipitation concentration will occur in May–June over that area. In addition, intense rainfall events and consecutive wet days will increase in frequency, which can expose the Gulf of Guinea and Savanna to flood events from June onwards. This confirms the results from Donat et al. (2016), which showed that the intensification of the hydrological cycle both in recent decades and in future projections will lead to an increased risk of flooding in dry regions as the climate warms. Groundnut, cassava, and maize are important crops for the Gulf of Guinea, especially for Nigeria, southern Mali, Benin, Ivory Coast, Burkina Faso, Ghana, and Senegal, while over the Savanna, the main crops are yam, millet, and sorghum. Therefore, to inform farmers about short-term coping and adaptation practices, scientists are encouraged to simulate crop models and to assess their uncertainties according to the shift in the times of the projected precipitation distribution and the increase in the soil temperature. Alternative strategies, such as constructing infrastructure or irrigation systems, could also be used to mitigate the impact of exposure.

Regional provisions and strategies that include all West African countries should be developed to meet the challenge of combating GHG production. The framework agreements must link and bind countries to ensure strict compliance with community-based adaptation measures. At the local level, moreover, each country will have to develop precautionary flood and drought warning systems to limit the loss of human life. Scientific community frameworks need to be developed at the local level to improve the seasonal prediction of rainfall models that must be updated frequently in order to generate reliable information. Better research findings are needed to increase knowledge of how information structures could be framed and used to reduce the power of parochial conflicting benefits and overcome inertia, apathy, and lack of political drive. Finally, communication systems geared toward achieving specific targets (e.g., to assist farmers) will have to be developed. Informing media platforms about climate sciences and adaptation strategy policies and discussions, for instance, would educate the public about the impacts of global warming, the importance of reducing GHG emissions, and the need for developing and implementing mitigation and adaptation strategies.

Discussion and Conclusion

It is widely demonstrated in the literature that West Africa is vulnerable to climate change due to its high climate variability and high reliance on rain-fed agriculture. It does not have an institutional capacity to respond and to adapt climate variability and climate change (Quenum et al. 2019). In order to reduce risks and suggest reliable adaptation strategies for predicted climate change, this study focused on determining the variability of precipitation both in the present time and under various future scenarios. The findings obtained with regard to the PCI showed that in West Africa the period summarizing the main rainfall activity is between May and September. For the historical period, simulations show over the Gulf of Guinea and the Savanna a uniform precipitation distribution, and in the Sahel both moderate and irregular distribution of precipitation. Under future scenarios, i.e., at all the GWLs (1.5 °C, 2.0 °C, 2.5 °C, and 3.0 °C in this study), the magnitude of the simulated PCIs over West Africa reduced and became close to PCI<10. This demonstrates that under the selected GWLs, the precipitation concentration of West African becomes more uniformly distributed except for some countries in the north-eastern areas (Niger and Chad), which are least dry at all four GWLs. To get further details on the temporal patterns of the precipitation concentration in the study domain, the PCP variable is computed, which reveals that the magnitude of the PCI increases northwardly materializing a south-north precipitation gradient. The highest precipitation concentrations during the control period occur in July–August and cover the Gulf of Guinea and the Savanna regions, whereas the Sahel records the peaks of its precipitation concentration in September. The PCP too changed in response to increasing GWLs. Therefore, the precipitation for the projected period becomes more concentrated in June–July over the Gulf of Guinea and the Savanna, and in August for the Sahel region. Thus, the rainfall concentration starts one month earlier in the future period compared to the historical (which is very important and needed information for the agricultural sector). Globally, it also noticed that the Savanna-Sahel region recorded a high magnitude of PCD. This means that the total yearly precipitation recorded in this region occurs in a short time, due to the WAM system, which is led by the back-and-forth movement of the Inter-Tropical Discontinuity (ITD) between south and north. Indeed, this movement creates an increased precipitation concentration in the Savanna-Sahel area, which is immediately followed by the southward movement of the ITD. This establishes a long time record of precipitation concentration, which is highlighted by the PCD values over this area. A significant reduction in the CDD is recorded over the north-east (i.e., Niger and Chad), and a slight increase in the number of CWD over the Sahel.

Additionally, based on the results from PCI and PCP, Niger and Chad are projected to experience more wet condition under increasing GWLs. The 5-day cumulative rainfall variable shows that the Gulf of Guinea is projected to experience more intense, very intense, and heavy rainfall events under increasing GWLs. All these results together show how much West Africa will be exposed to higher variability of climate change and also to future heavier rainfall and wet conditions.

To cope with such changes, reduce loss of life, and better manage the impact on inhabitants of this region, some adaptation strategies are necessary under continual climate change. Two types of strategies are required: a regional framework agreement and local coping strategies. The regional agreement is very important because it forces each stakeholder (country) to respect the framework agreement; this is mostly in line with mitigation of climate change at the regional scale. Local strategies are important in two ways: Firstly, local strategies should enable West African countries to respect the regional framework agreement. Secondly, they should encourage countries to look for and develop adequate adaptation possibilities, by responding to the contributions of scientists and decision makers, which represent an important factor in development. They are the ones who have to provide reliable information to the population, and particularly to farmers, for better management of crops in order to ensure food security.

In summary, this study has investigated precipitation variability and change in West Africa under increasing GWLs, and identified an earlier onset of rainfall, especially over the Gulf of Guinea. It has found that the variability of CDDs and CWDs under increasing GWLs is considerable, and that even the intensity of rainfall has increased. Such significant information is useful for farmers and decision makers to ensure the survival and prosperity of the population.

Acknowledgments This work was funded by the German Federal Ministry of Education and Research (BMBF) through the West African Science Service Center on Climate Change and Adapted Land Use (WASCAL), which provided a research fund to Gandome Mayeul L.D. Quenum. The Centre for High-Performance Computing (CHPC, South Africa) provided the computing facility used for the study.

References

Abiodun BJ, Makhanya N, Petja B, Abatan AA, Oguntunde PG (2019) Future projection of droughts over major river basins in Southern Africa at specific global warming levels. Theor Appl Climatol 137(3–4):1785–1799. https://doi.org/10.1007/s00704-018-2693-0

Barron J, Rockström J, Gichuki F, Hatibu N (2003) Dry spell analysis and maize yields for two semi-arid locations in east Africa. Agric For Meteorol, Elsevier 117:23–37. https://doi.org/10.1016/S0168-1923(03)00037-6

Chou C, Lan C-W (2011) Changes in the annual range of precipitation under global warming. J Clim 25:222–235. https://doi.org/10.1175/JCLI-D-11-00097.1

De Luis M, González-Hidalgo JC, Brunetti M, Longares LA (2011) Precipitation concentration changes in Spain 1946–2005. Nat Hazards Earth Syst Sci 11(5):1259–1265. https://doi.org/10.5194/nhess-11-1259-2011

De Perthuis C, Hallegatte S, Lecocq F (2010) Économie de l' adaptation au changement climatique. Conseil économique pour le développement durable, Ed, Paris

Déqué M, Calmanti S, Bøssing O, Dell A, Fox C, Haensler A, … Teichmann C (2017) A multi-model climate response over tropical Africa at + 2°C. Climate Ser (Elsevier) 7:87–95. https://doi.org/10.1016/j.cliser.2016.06.002

Diasso U, Abiodun BJ (2015) Drought modes in West Africa and how well CORDEX RCMs simulate them. Theor Appl Climatol 128(1–2):223–240. https://doi.org/10.1007/s00704-015-1705-6

Donat MG, Lowry AL, Alexander LV, Gorman PAO, Maher N (2016) More extreme precipitation in the world' s dry and wet regions. Nat Clim Chang 6(May). https://doi.org/10.1038/NCLIMATE2941

Easterling DR, Meehl GA, Parmesan C, Changnon SA, Karl TR, Mearns LO (2000) Climate extremes: observations, modeling, and impacts. Sci Compass 289:2068–2074. https://doi.org/10.1126/science.289.5487.2068

Egbebiyi TS (2016) Future changes in extreme rainfall events and African easterly waves over west Africa. University of Cape Town, Cape Town

Ezenwaji EE, Nzoiwu CP, Chima GN (2017) Analysis of precipitation concentration index (PCI) for Awka Urban area. Hydrol Current Res 8(4):4–9. https://doi.org/10.4172/2157-7587.1000287

FAO, IFAD, WFP. (2015) The state of food insecurity in the world 2015. Meeting the 2015 international hunger targets: taking stock of uneven progress. (FAO, ed). Rome

Frich P, Alexander LV, Della-Marta P, Gleason B, Haylock M, Tank Klein AMG, Peterson T (2002) Observed coherent changes in climatic extremes during the second half of the twentieth century. Clim Res 19(3):193–212. https://doi.org/10.3354/cr019193

Füssel H-M (2007) Adaptation planning for climate change: concepts, assessment approaches, and key lessons. Sustain Sci 2:265–275. https://doi.org/10.1007/s11625-007-0032-y

Gao Y, Zhu X, Yu G, He N, Wang Q, Tian J (2014) Agricultural and forest meteorology water use efficiency threshold for terrestrial ecosystem carbon sequestration in China under afforestation. Agric For Meteorol 195–196:32–37. https://doi.org/10.1016/j.agrformet.2014.04.010

Haensler A, Saeed F, Jacob D (2013) Assessing the robustness of projected precipitation changes over central Africa on the basis of a multitude of global and regional climate projections. Clim Chang 121:349–363. https://doi.org/10.1007/s10584-013-0863-8

IPCC (2001) Climate change 2001: synthesis report. A contribution of working groups I, II, and III to the third assessment report of the integovernmental panel on climate change (eds: Watson RT, The Core Writing Team). Cambridge University Press, Cambridge/New York

IPCC (2014) Climate change 2014: synthesis report. Contribution of working groups I, II and III to the fifth assessment report of the intergovernmental panel on climate change (eds: Core Writing Team, Pachauri RK, Meyer LA). IPCC, Geneva, p 151

Klutse NAB, Ajayi V, Gbobaniyi EO, Egbebiyi TS, Kouadio K, Nkrumah F, … Dosio A (2018) Potential impact of 1.5°C and 2°C global warming on consecutive dry and wet days over West Africa. Environ Res Lett. https://doi.org/10.1088/1748-9326/aab37b

Kpadonou RA, Adegbola PY, Tovignan SD (2012) Local knowledge and adaptation to climate change in Ouémé climate change is a major threat to sustainable development in developing countries. According to IPCC (2007), poor communities particularly in sub-Saharan Africa will be most vulnerable because. Afr Crop Sci J 20(Supplement S2–20):181–192

Kumi N, Abiodun BJ (2018) Potential impacts of 1.5°C and 2°C global warming on rainfall onset, cessation and length of rainy season in West Africa. Environ Res Lett 13(5):12–13. https://doi.org/10.1088/1748-9326/aab89e

Li X, Jiang F, Li L, Wang G (2011) Spatial and temporal variability of precipitation concentration index, concentration degree and concentration period Xinjiang, China. Int J Climatol 31 (11):1679–1693. https://doi.org/10.1002/joc.2181

Lobell DB (2014) Climate change adaptation in crop production : beware of illusions. Glob Food Sec:1–5. https://doi.org/10.1016/j.gfs.2014.05.002

Locatelli B (2011) Synergies between adaptation and mitigation in a nutshell. CIFOR

Maúre G, Pinto I, Ndebele-Murisa M, Muthige M, Lennard C, Nikulin G, . . . Nicholson SE (2018) The southern African climate under 1.5°C and 2°C of global warming as simulated by CORDEX regional climate models. Environ Res Lett 13(6):065002. https://doi.org/10.1088/1748-9326/aab190

Moss R, Babiker M, Brinkman S, Calvo E, Carter T, Edmonds J, . . . Zurek M (2008) Towards new scenarios for analysis of emissions, climate change, impacts, and response strategies. IPCC expert meeting report. Intergovernmental Panel on Climate Change, Geneva/Noordwijkerhout

Nikulin G, Jones C, Giorgi F, Asrar G, Büchner M, Cerezo-Mota R, . . . Sushama L (2012) Precipitation climatology in an ensemble of CORDEX-Africa regional climate simulations. J Climate 25(18):6057–6078. https://doi.org/10.1175/JCLI-D-11-00375.1

Nikulin G, Lennard C, Dosio A, Kjellström E, Chen Y, Hansler A, . . . Somot S (2018) The effects of 1.5 and 2 degrees of global warming on Africa in the CORDEX ensemble. Environ Res Lett 13 (6). https://doi.org/10.1088/1748-9326/aab1b1

Oliver JE (1980) Monthly precipitation distribution: a comparative index. Prof Geogr 32(3):300–309. https://doi.org/10.1111/j.0033-0124.1980.00300.x

Orlowsky B, Seneviratne SI (2012) Global changes in extreme events: regional and seasonal dimension. Clim Chang 110:669–696. https://doi.org/10.1007/s10584-011-0122-9

Quenum GMLD, Klutse NAB, Dieng D, Laux P, Arnault J, Kodja JD, Oguntunde PG (2019) Identification of potential drought areas in West Africa under climate change and variability. Earth Syst Environ 3(3):429–444. https://doi.org/10.1007/s41748-019-00133-w

Rowhani P, Lobell DB, Linderman M, Ramankutty N (2011) Agricultural and forest meteorology climate variability and crop production in Tanzania. Agric For Meteorol 151(4):449–460. https://doi.org/10.1016/j.agrformet.2010.12.002

Saba A, Gerrard MB, Lobell DB, Saba A, Biasutti M, Gerrard MB, Lobell DB (2013) Getting ahead of the curve : supporting adaptation to long-term climate change and short-term climate variability alike. CCLR 7. Retrieved from https://scholarship.law.columbia.edu/faculty_scholarship/563

Sheffield J, Wood EF, Chaney N, Guan K, Sadri S, Yuan X, . . . Ogallo L (2014) A drought monitoring and forecasting system for sub-sahara african water resources and food security. Am Meteorol Soc 95(June), 861–882. https://doi.org/10.1175/BAMS-D-12-00124.1

Shi P, Wu M, Qu S, Jiang P, Qiao X, Chen X, . . . Zhang Z (2015) Spatial distribution and temporal trends in precipitation concentration indices for the Southwest China. Water Resour Manag 29 (11):3941–3955. https://doi.org/10.1007/s11269-015-1038-3

Sultan B, Gaetani M (2016) Agriculture in West Africa in the twenty-first century: climate change and impacts scenarios, and potential for adaptation. Front Plant Sci 7(August):1–20. https://doi.org/10.3389/fpls.2016.01262

Sylla MB, Gaye AT, Jenkins GS (2012) On the fine-scale topography regulating changes in atmospheric hydrological cycle and extreme rainfall over West Africa in a regional climate model projections. Int J Geophys 2012:1. https://doi.org/10.1155/2012/981649

UNFCCC (2010) Framework convention on climate change conference of the parties report of the conference of the parties on its sixteenth session, held in Cancun from 29 November to Part Two: action taken by the conference of the parties at its sixteenth session The Ca, (March), 1–31

Van Vuuren DP, Edmonds J, Kainuma M, Riahi K, Nakicenovic N, Smith SJ, Rose SK (2011) The representative concentration pathways : an overview. Clim Chang 109:5–31. https://doi.org/10.1007/s10584-011-0148-z

Wang X, Zhang J, Yang Z (2013) Historic water consumptions and future management strategies for Haihe River basin of Northern China. Mitig Adapt Strateg Glob Chang 20:371. https://doi.org/10.1007/s11027-013-9496-5

Watkiss P, Brown S, Christensen OB, Maule CF, Hinkel J, Jacob D, . . . Nicholls R (2015) Effects of 2 C Warming-IMPACT2C modelling results: climate change and sea-level rise from a 2 C climate. IMPACT2C Policy Brief, 2, 36

Wiebe K, Sulser TB, Croz DM, Rosegrant MW (2017) The effects of climate change on agriculture and food security in Africa. International Food Policy Research Institute (IFPRI), Washington, DC

Zhang L, Qian Y (2003) Annual distribution features of precipitation in china and their interannual variations. Acta Meteor Sin 17(2):146

Zhang Q, Liu CX, Gemmer M, Chen YD, Liu C (2009) Changing properties of precipitation concentration in the Pearl River basin. China:377–385. https://doi.org/10.1007/s00477-008-0225-7

5

Climate Change Impact on Climate Extremes and Adaptation Strategies in the Vea Catchment, Ghana

Isaac Larbi, Clement Nyamekye, Fabien C. C. Hountondji, Gloria C. Okafor and Peter Rock Ebo Odoom

Contents

I. Larbi (✉)
Climate Change and Water Resources, West African Science Service Centre on Climate Change and Adapted Land Use (WASCAL), Université d'Abomey-Calavi, Cotonou, Benin

School of Sustainable Development, University of Environment and Sustainable Development, Somanya, Ghana

Department of Civil Engineering, Faculty of Engineering, Koforidua Technical University, Koforidua, Ghana
e-mail: larbi.i@edu.wascal.com

C. Nyamekye
Department of Civil Engineering, Faculty of Engineering, Koforidua Technical University, Koforidua, Ghana

F. C. C. Hountondji
Faculté d'Agronomie, University of Parakou, Parakou, Benin

G. C. Okafor
Department of Civil Engineering, Nigeria Maritime University, Delta–State, Nigeria

P. R. E. Odoom
Climate Change and Water Resources, West African Science Service Centre on Climate Change and Adapted Land Use (WASCAL), Université d'Abomey-Calavi, Cotonou, Benin

Abstract

Climate change impact on rainfall and temperature extreme indices in the Vea catchment was analyzed using observation and an ensemble mean of bias-corrected regional climate models datasets for Representative Concentration Pathway (RCP 4.5) scenario. Rainfall extreme indices such as annual total wet-day precipitation (PRCPTOT), extremely wet days (R99P), consecutive wet days (CWD), consecutive dry days (CDD), and temperature indices such as warmest day (TXx) and warmest night (TNx) from the Expert Team on Climate Change Detection Monitoring Indices (ETCCDMI) were computed for both the historical (1986–2016) and future (2020–2049) period using the RClimdex. The parametric ordinary least square (OLS) regression approach was used to detect trends in the time series of climate change and extreme indices. The results show an increase in mean annual temperature at the rate of 0.02 °C/year and a variability in rainfall at the catchment, under RCP 4.5 scenario. The warmest day and warmest night were projected to increase by 0.8 °C and 0.3 °C, respectively, in the future relative to the historical period. The intensity (e.g., R99p) and frequency (e.g., CDD) of extreme rainfall indices were projected to increase by 29 mm and 26 days, respectively, in the future. This is an indication of the vulnerability of the catchment to the risk of climate disasters (e.g., floods and drought). Adaptation strategies such as early warning systems, availability of climate information, and flood control measures are recommended to reduce the vulnerability of the people to the risk of the projected impact of climate extreme in the future.

Keywords

Climate change · Climate models · Early warning systems · Climate extreme indices · Vea catchment

Introduction

In the last century, global surface temperature has increased by 0.71 °C with significant warming observed in many regions (Trenberth et al. 2007). This increase in global surface temperature reveals more warming of land than the oceans and has diverse consequences on water resources and the livelihood of people. In view of this, much attention has been given to climate change and extremes over the past decades (New et al. 2006; Larbi et al. 2018; Adeyeri et al. 2019) due to its intense impact on natural and human systems. At the study region, the high variability in rainfall has also resulted in higher frequency of flood and drought events, putting the socioeconomic activities at risk (Awotwi et al. 2015).

The impact of climate change on climate extremes and other sectors require tools for its assessment, and the most common approach for the assessment is the use of climate model outputs. Though Global Circulation Models (GCMs) have been widely used for various climate change studies (Minville et al. 2009; Chen et al. 2012; Panday et al. 2015), it is not suitable for analyzing the projections on a regional or local scale due to its coarse resolution nature (Xu 1999; Maraun et al. 2010; Seager and Vecchi 2010). In West Africa regions, GCMs are unable to represent key features of the West African monsoon (WAM) such as the Africa Easterly Waves (AEWs), Africa Easterly Jets (AEJ), and Tropical Easterly Jets (TEJ) (Sylla et al. 2013). Regional climate models (RCMs) are therefore used in climate change studies because of its ability to accurately take into account orography, thus providing a better reproduction of regional climates (Sylla et al. 2013).

Extreme climatic events such as heat waves, floods, and droughts affect lives and livelihoods of people over varied regions (Alexander et al. 2006). Floods and droughts generate socioeconomic effects such as hunger, habitat, and infrastructure destruction. Over 40% of the populace has encountered hunger situations, that resulted from failure of one-third of crop produced when compared to 2010 data, leading to humanitarian appeals (Sarr 2012; FAO 2013). These and other ensuing impacts have necessitated the analysis of changes in future climate extremes necessary in reducing potential social, economic, and ecological consequences (Amuzu et al. 2018; Bohle et al. 2018). This need has necessitated studies in the analysis of past trends and the prediction of the occurrence of extreme climate events on rainfall and temperature using different approaches. Several studies have therefore achieved this by the use of RCMs. This model has been frequently used for climate change impact studies (Zhang et al. 2008) and future extreme climate projections analysis (Tomassini and Jacob 2009).

In West Africa, M'Po et al. (2017) assessed climate change impact on rainfall extremes over the Ouémé River Basin in Benin using RCM. The results showed a decrease in extremely heavy precipitation and annual total wet-day precipitation in most stations for the future 2015–2050. The Vea catchment in Ghana is seen as one of the vulnerable areas to climate change and is faced with issues of climate shocks such as flood and drought. This has devastating effects on various sectors of the economy particularly water resources and agriculture. However, information regarding climate change and climate extremes which are vital to improve societal awareness and preparedness is limited in the Vea catchment. To this end, understanding trends and variations of historical and future climatic variables is pertinent for the future development and sustainable water resources management in the catchment (Ogountundé et al. 2006). This chapter assesses the impact of climate change on rainfall and temperature extreme indices in the Vea catchment in the Upper East Region of Ghana. This is based on the Representative Concentration Pathways (RCP 4.5) scenario, which would recommend adaptation strategies required at the community level to reduce the vulnerability of the people to the risk of climate change and extreme events in the future.

Description of the Study Area and Methodology

This section describes the area where the study was conducted, the data used, and the methodology applied to achieve the aim of the study. The Vea catchment, one of the sub-catchments within the White Volta basin (WVB), is located between latitudes 10° 30′N–11° 08′N and longitudes 1° 15′W–0° 50′E (Fig. 1). It has an area of about 305 km^2 and covers mainly the Bongo and Bolgatanga districts in the Upper East Region of Ghana, with a small portion over the south-central part of Burkina Faso. The climate of the catchment is controlled by the movement of the Inter-tropical Discontinuity (ITD) over the land that dominates the climate of the entire West African region (Obuobie 2008). Located in a semiarid agroclimatic zone, the catchment covers three agroecological zones: the Savanna and Guinea Savanna zones in Ghana, and north Sudanian Savanna zone in Burkina Faso (Forkuor 2014). It is characterized by a unimodal rainfall regime from April/May to October with a mean annual rainfall of 957 mm which normally peaks in August (Larbi et al. 2018). The temperature is uniformly high with a mean value of 28.5 °C while potential evapotranspiration in the catchment exceed monthly rainfall for most part of the year, except the three wettest months of July, August, and September (Limantol et al. 2016). It is characterized by fairly low relief with elevation ranging between 89 m and 317 m (Fig. 1) and mainly dominated by cropland followed by grassland interspersed with shrubs and trees and woodland (closed/open).

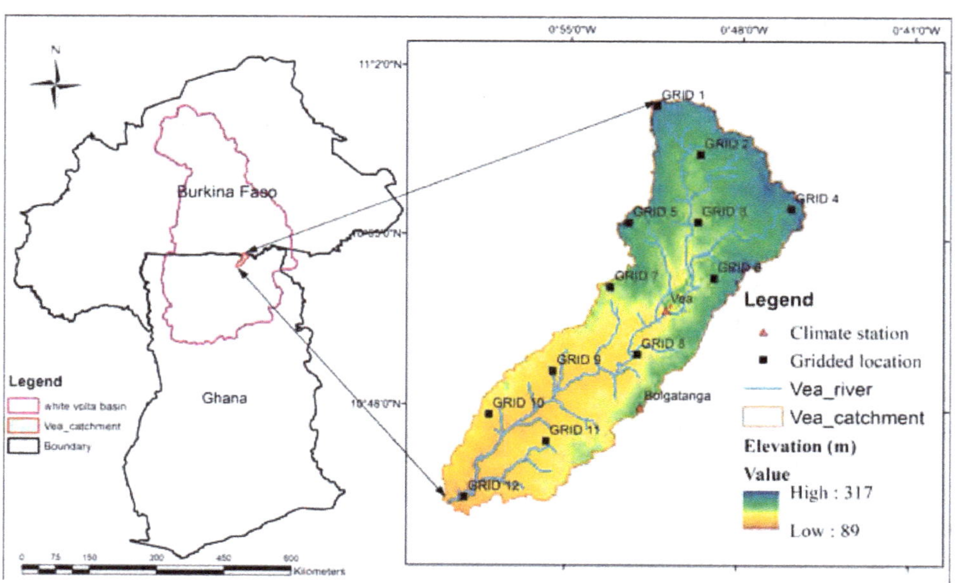

Fig. 1 Map of the Vea catchment showing the station and gridded precipitation locations within a 12 km grid

Observation and Climate Change Scenario Datasets

Observed daily rainfall, maximum and minimum temperature within the Vea catchment from 1986 to 2016 for two climate stations (Vea and Bolgatanga) were obtained from the West African Science Service Centre on Climate Change and Adapted Land Use (WASCAL) research center and Ghana Meteorological Agency. Additional 12 (Fig. 1) gridded daily precipitation data at 0.05° spatial resolution from Climate Hazards Group InfraRed Precipitation with Station data (CHIRPS) were also downloaded from http://chg.geog.ucsb.edu/data/chirps (Funk et al. 2015). CHIRPS incorporate satellite imagery with in-situ station data to create gridded rainfall time series. The performance of the CHIRPS data in reproducing the climatology of the Vea catchment has been evaluated in previous research (Larbi et al. 2018). According to Larbi et al. (2018), the CHIRPS data is able to reproduce well both the seasonal and annual rainfall pattern of the Vea catchment, and the validation resulted in a very high correlation coefficient ($r = 0.99$) and a Nash–Sutcliffe efficiency of 0.9.

Four RCMs datasets (Table 1) within the CORDEX-Africa experiment and Weather Research and Forecasting (WRF) models were obtained from both CORDEX-Africa website and WASCAL geoportal (Heinzeller et al. 2016), respectively. The CORDEX-Africa RCMs have been selected based on its ability to simulate the basic climatological features in West Africa (Kim et al. 2013). The WRF simulations were produced by the Karlsruhe Institute of Technology, Institute of Meteorology and Climate Research, Atmospheric Environmental Research (KIT/IMK-IFU), Germany, and the West African Science Service Centre on Climate Change and Adapted Land Use (WASCAL) regional climate simulations. The WRF (WRF-H and WRF-G) simulations were downscaled from two GCMs namely: the General Fluid Dynamics Laboratory Earth System Model (GFDL-ESM 2 M) and the Hadley Global Environment Model (HadGEM2-ES) using the Weather Research and Forecasting Model (WRFv3.5.1). A detailed technical description and parameterization of the WRF models have been reported by Heinzeller et al. (2017). The CORDEX-Africa RCMs (REMO2009 and KNMI-RACMO22T) and the WRF models were selected due to its remarkable skills in reproducing the climatology of the Vea catchment for the period 1981–2005 (Larbi et al. 2020). The WRF models skills have also demonstrated to do well in terms of rainfall distribution in a study conducted within West

Table 1 Description of the CORDEX-Africa and WRF models whose ensemble mean was used

GCMs	RCMs	Institution	Resolution
ICHEC-EC-EARTH	REMO2009	Max Planck Institute – Computational methods in systems and control theory (MPI-CSC), Germany	50 km
ICHEC-EC-EARTH	KNMI-RACMO22T	Koninklijk Nederlands Meteorologisch Instituut (KNMI)	50 km
HadGEM2-ES	WRF-H	WASCAL / KIT/IMK-IFU	12 km
GFDL-ESM2M	WRF-G	WASCAL / KIT IMK-IFU	12 km

Africa by Bessah et al. (2018). Details on the RCMs evaluation and bias-correction can be found in Larbi et al. (2020). The ensemble mean of the bias-corrected RCMs output used consist of daily rainfall and temperature for the future (2020–2049) under Representative Concentration Pathway (RCP4.5) climate change scenario.

Climate Extreme Indices Analysis

Several climate extreme indices have been developed by the Expert Team on Climate Change Detection Monitoring Indices (ETCCDMI) for understanding climate extremes and are widely used in several regions (Mouhamed et al. 2013; Soro et al. 2016; M'Po et al. 2017). Six of the indices were adopted. The selected rainfall and temperature extreme indices (Table 2) were computed for each climate location for both the historical (1986–2016) and future (2020–2049) period on the interface of R software using RClimdex. The historical extreme indices was computed using climate observation data while the future climate extreme indices were computed based on the ensemble mean of the RCMs data for the RCP 4.5 climate change scenario. The computed rainfall and temperature extremes indices at the annual scale were analyzed at spatial and temporal scale using the inverse distance weighting (IDW) technique (Feng-Wen and Chen-Wuing 2012). Trends in the climate extreme indices time series were computed based on the parametric linear regression approach (M'Po et al. 2017).

Presentation of Results and Discussion

Annual Rainfall and Temperature Projections and Trends

The results for the projections of mean annual rainfall and temperature in the Vea catchment by the ensemble mean of the RCMs under the RCP4.5 scenario is shown in Fig. 2 and Table 3. The ensemble mean of the RCMs shows an increase in mean annual rainfall and temperature by 22.6 mm (2.4%) and 1.3 °C, respectively, for the

Table 2 Climate extreme indices

Indices	Descriptive name	Definition	Units
PRCPTOT	Annual total wet-day precipitation	Annual total precipitation from days ≥1 mm	mm
R99p	Extremely wet days	Annual total precipitation on the days when daily PRCP >99th percentile	mm
CWD	Consecutive wet days	Maximum number of consecutive days with PRCP ≥1 mm	days
CDD	Consecutive dry days	Maximum number of consecutive days with PRCP<1 mm	days
TXx	Warmest day	Annual maximum value of the daily max temperature	°C
TNx	Warmest night	Annual maximum value of daily min temperature	°C

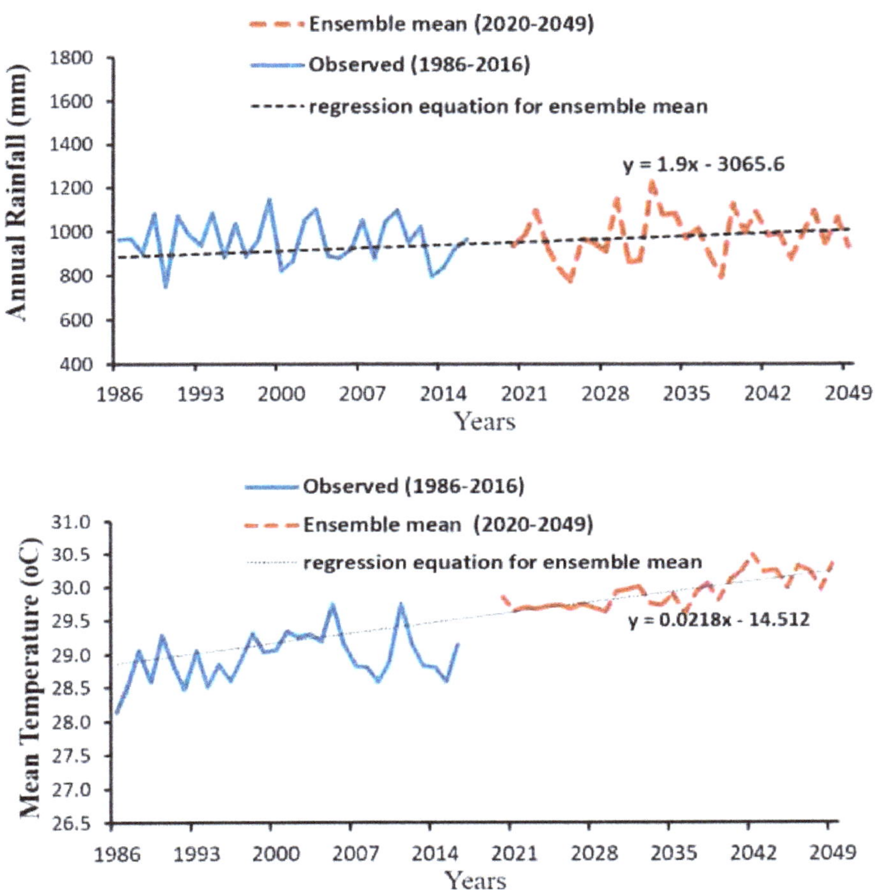

Fig. 2 Mean annual rainfall and temperature projection for the Vea catchment in the future (2020–2049) under RCP 4.5 scenario relative to the 1986–2016 period

period 2020–2049 relative to the baseline (1986–2016) period. An increasing trend in the future annual rainfall and temperature was found to be at the rate of 1.9 mm/year and 0.02 °C/year.

Trends and Projected Changes in Extreme Rainfall Indices

The results for the extreme rainfall indices (PRCPTOT, CWD, CDD, and R99p) projected by the ensemble mean of the RCMs under RCP4.5 scenario relative to the baseline (1986–2016) period are shown in Table 4 and Fig. 3. The annual total precipitation (PRCPTOT) is projected to increase from 951.9 mm to 952.1 mm (Table 4) at an annual rate of 2.2 mm/year (Fig. 3a). Extremely wet days (R99p) is projected to increase by 29.1 mm; however, a decreasing trend at a rate of 0.77 mm/year is projected in the future 2020–2049 (Fig. 3d). Unlike consecutive dry days

Table 3 Mean annual rainfall and temperature projections for the Vea catchment relative to the baseline period

Variables	Baseline (1986–2016)	RCP4.5 (2020–2049)	Change
Tmax (°C)	34.6	35.9	1.3
Tmin (°C)	22.6	24.0	1.4
Tmean(°C)	28.6	29.9	1.3
Rainfall (mm)	958.5	981.1	2.4%

Table 4 Projected changes in mean annual climate extreme indices at the Vea catchment

Climate extreme indices	Baseline (1986–2016)	RCP4.5 (2020–2049)	Change
PRCPTOT (mm)	951.9	952.1	0.2
R99p (mm)	46.6	75.7	29
CWD (days)	5	4	−1
CDD (days)	105	131	26
TXx (°C)	41.5	42.3	0.8
TNx (°C)	28.9	29.2	0.3

(CDD) which is projected to increase from 105 to 131 days, consecutive wet days (CWD) is projected to decrease by 1 day in the future 2020–2049 (Table 4). The spatial distribution of the extreme rainfall indices (Fig. 4) indicates a projected increase in CDD in most part of the catchment, with the exception of the center parts of the catchment that shows a decrease of 5 days (Fig. 4c). The PRCPTOT is projected to decrease in most parts of the catchment except the extreme north and south which shows an increase of 28.7 mm (Fig. 4f). R99p is projected to increase in all parts of the catchment especially in the northern and southern parts and will range from 27.7 to 35.7 mm (Fig. 4L). Unlike R99p, CWD is projected to decrease in the entire catchment in the range of −3.4 to −0.4 days (Fig. 4I).

Temperature Extreme Indices Projections and Trends

The time series of the projected extreme temperature indices (TXx and TNx) and its spatial distribution are shown in Figs. 5 and 6, respectively. As shown in Table 4, the warmest day (TXx) is projected to increase by 0.8 °C from 41.5 °C to 42.3 °C. The warmest night (TNx) is also projected to increase from 28.9 °C to 29.2 °C. The future trend analysis of the extreme temperature indices shows an increasing trend in both TXx and TNx at a rate of 0.04 and 0.01 °C/year, respectively. At spatial scale, TXx is projected to increase in the entire catchment for the period 2020–2049, with a higher increase in the southern part of the catchment in the range of 0.79 to 1.29 °C. The TNx is also projected to increase in the range of 0.0 °C to 0.48 °C in the greater part of the Vea catchment.

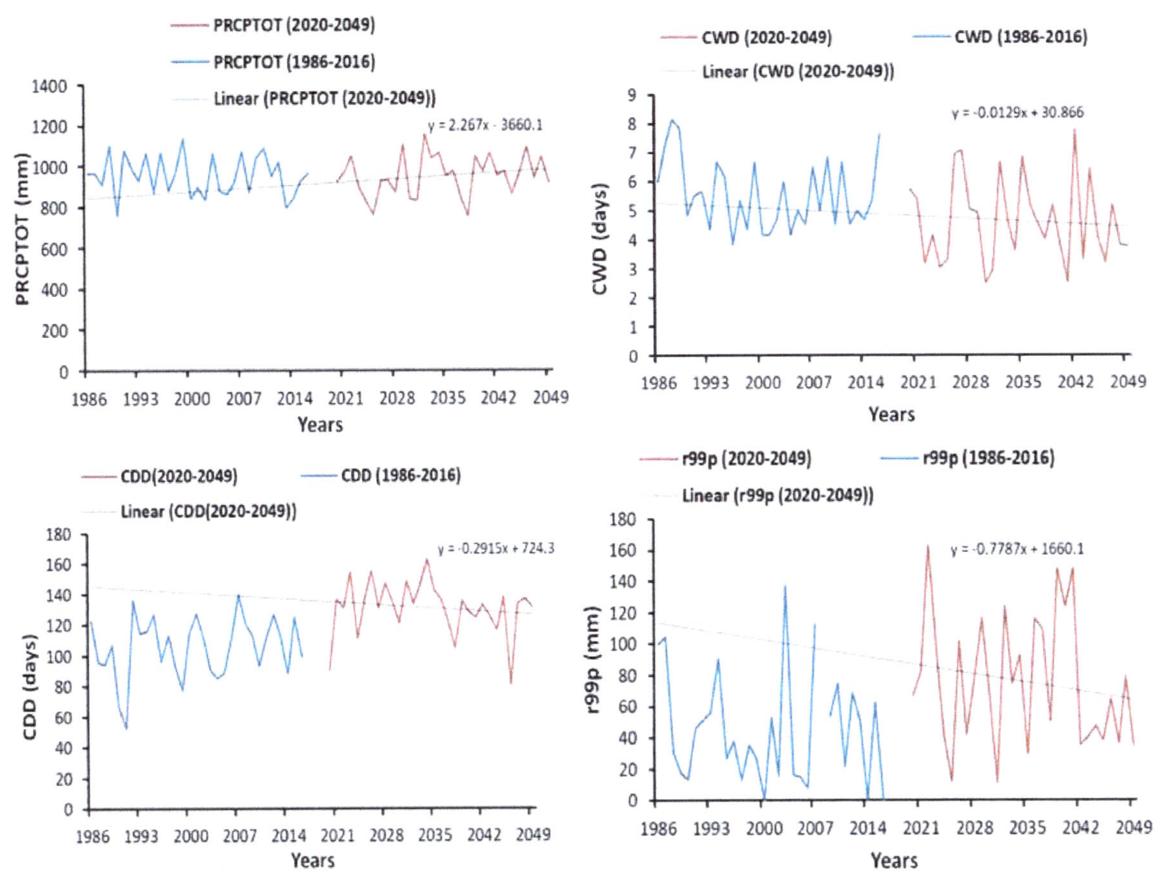

Fig. 3 Trend analysis of extreme rainfall indices: (**a**) PRCPTOT, (**b**) CWD, (**c**) CDD, and (**d**) r99p at the Vea catchment for the historical (1986–2016) and future (2020–2049) period

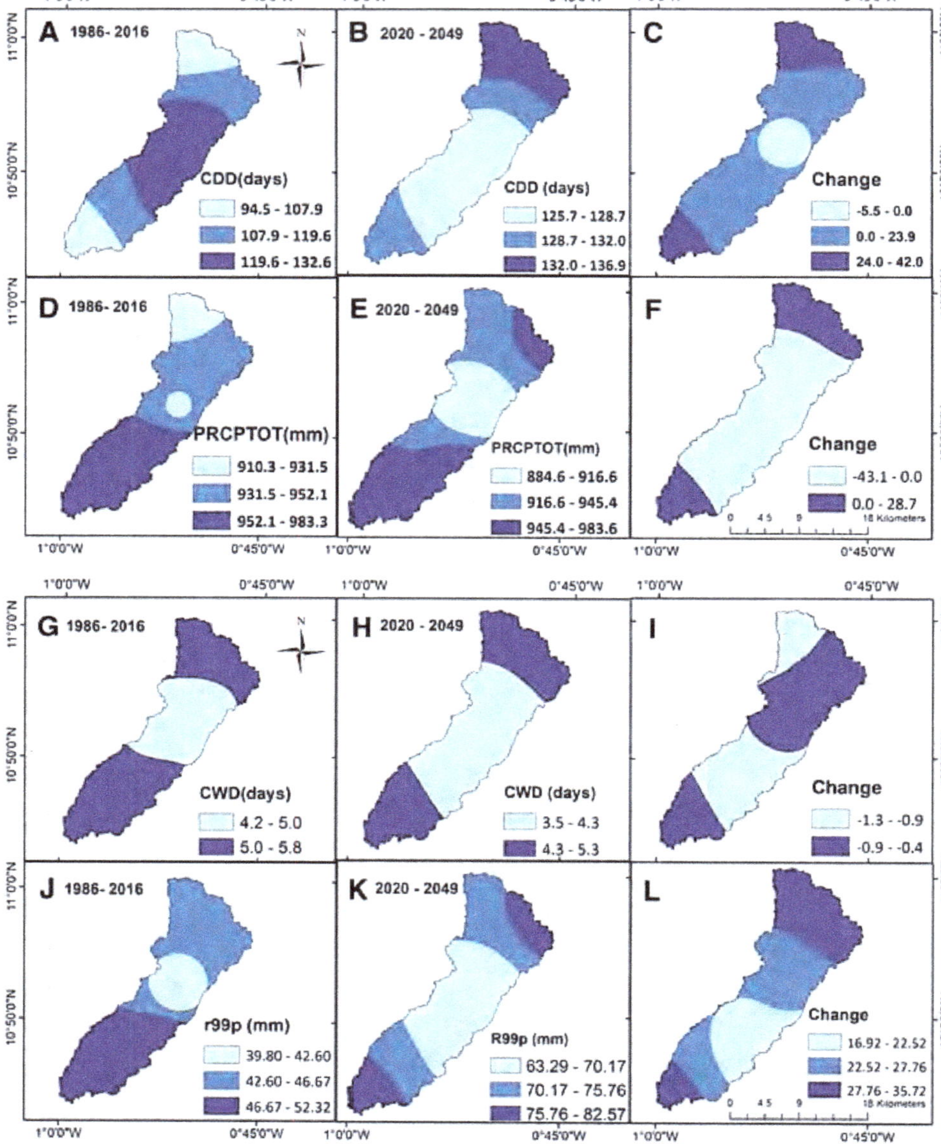

Fig. 4 Spatial distribution of the projected changes in the extreme rainfall indices: (**a**) CDD, (**b**) PRCPTOT, (**c**) CWD, and (**d**) r99p in the Vea catchment

Climate Change and Climate Extreme Impacts at the Study Region

The agroclimatic zone of the Vea catchment is characterized as semiarid region in West Africa and can influence growth when managed efficiently. With the expected increase in precipitation and temperature, there is a likelihood of potential risk to

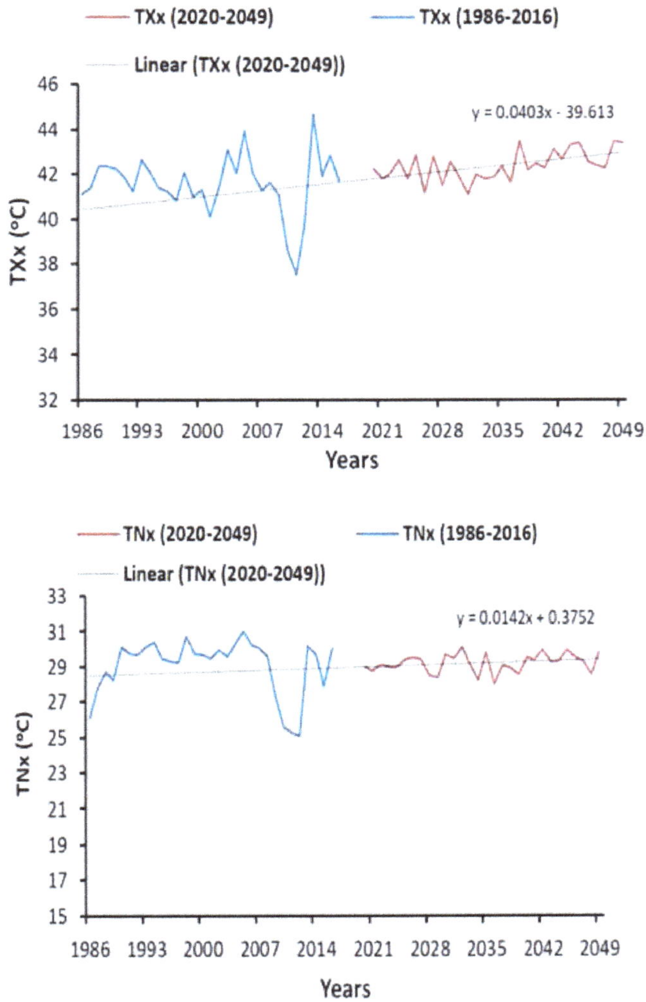

Fig. 5 Trend analysis of extreme temperature indices for TXx and TNx in the Vea catchment for the historical and future period

natural disasters and destruction of infrastructure which would have environmental, economic, and societal implications (Gobiet et al. 2014). The climate analysis performed on the daily rainfall and temperature series for the historical and future period provides evidence that the Vea catchment would be affected by climate change and warm extremes based on the indices analyzed with warming more pronounced during the day than the night. Although in Ghana climate change projections vary considerably among different models, there is a consensus on the evidence of increasing temperature across the country. The mean annual temperature at the Vea catchment is projected to increase by 1.3 °C for the period 2020–2049. This finding is in line with observations reported by other climate change studies (e. g., McSweeney et al. 2010a; De Pinto et al. 2012). Additionally, a study by De Pinto et al. (2012) reported a projected increase in the mean annual temperature of Ghana

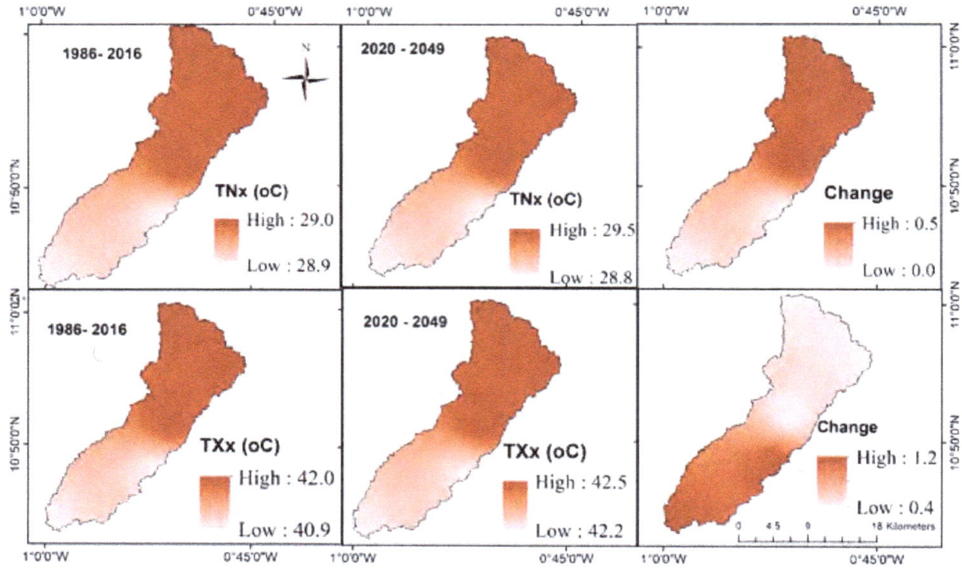

Fig. 6 Spatial distribution of the projected changes in the TNx and TXx temperature indices relative to the baseline (1986–2016) period

in the range of 1.0–3.0 °C by 2060 and by 1.5–5.2 ° C by 2090, with more rapid increases expected in the northern sector of the country. Similar increase in temperature of 0.6 °C, 2.0 °C, and 3.9 °C by the year 2020, 2050, and 2080, respectively, in Ghana has also be projected by the Ministry of Environment Science, Technology and Innovation of Ghana [MESTI] (2013). These projections in temperature are noticed to be across all regions in the country with a rapid increase in the northern sector (World Bank 2010). However, rainfall projections in Ghana are characterized with greater uncertainties with some models predicting an increase while others a decrease (McSweeney et al. 2010b).

In terms of climate extremes, the study region has witnessed periodic extreme climate shocks and disasters such as high temperature, highly variable and erratic rainfall, and long spells of droughts (Kanlisi and Arkum 2013). According to EPA (2007), such extreme weather events are expected to intensify in the region by 2080. The ensemble mean of the four climate models projected a nonsignificant increase in the mean annual rainfall; however, there exist spatial variations in the annual total rainfall (PRCPTOT), CWD, and CDD with some part of the Vea catchment indicating an increase while other parts indicate a decrease in the future period. These projections in climate change and extremes would have serious socioeconomic impacts on rural farmers in Ghana whose livelihoods depend largely on rainfall (Abeygunawardena et al. 2003; Fosu-Mensah et al. 2012). The changes in climate would affect agriculture and other sectors such as water resources, especially in the Upper East Region of Ghana where the Vea catchment is located (Gyasi et al. 2006). The projected warming trend at the Vea catchment lead to a higher evaporative demand from water bodies and irrigated crops in the future. Limantol et al. (2016)

indicated that temperature and evapotranspiration have been on the increase in the Vea catchment. And according to Ofori-Sarpong (2001), the increase in evapotranspiration could lead to soil moisture deficit and poor crop yield, particularly when there is no corresponding significant increase in rainfall. Also, evapotranspiration propelled by temperature will majorly affect agriculture through droughts and soil moisture deficits (Limantol et al. 2016).

Existing and Recommended Adaptation Strategies to Address Poverty and Ensure Sustainable Livelihoods in the Catchment

In sub-Saharan Africa, adapting to the impact of climate change on agriculture and other sectors of the economy has become a major concern to various stakeholders with special emphasis on improving adaptive capacity (Apata 2011). The projected climate change and extremes at the Vea catchment underscore the need to put in place appropriate adaptation measures to foster resilience to climate change in order to enhance water security and livelihood of the people within the area. In the Vea catchment, Limantol et al. (2016) examined farmers' perceptions and adaptation practices to climate change and variability, and noticed that farmers are aware of climate change and are adopting strategies to cope with the effects. At the catchment, some farmers use rain-fed practices to adjust to climate variability by varying crop types via rotation without fertilizer, while others employ irrigation practices to offset climate variability with a greater use of fertilizer application. According to Limantol et al. (2016), farmers identified irrigation development (access to water), followed by access to credit and health services as the most urgent needs for adaption to climatic change in the Vea catchment. Similarly, Kumasi et al. (2017) also reported that, the farmers in the Upper East Region of Ghana mostly practice crop selection, changes to planting dates and density, communal pooling (e.g., labor or income sharing across households), and traditional irrigation techniques as adaptation strategies to climate change. Additionally, Lawson et al. (2019) investigated the intersectional perceptions and adaptation strategies of women farmer in dealing with climate change in semiarid regions of Ghana. It was revealed that women farmers perceived the occurrence of climate change in the community and have incorporated some adaptation measures to counter its impacts. These adaptation strategies include changing planting dates, mixed farming, intercropping, planting early maturing varieties, composting, and seeking off-farm activities.

In Ghana, national climate change adaptation strategy has been developed and various adaptation programs and strategies have been outlined to combat climate change. Some of these strategies include: (i) increasing resilience to climate change impacts: identifying and enhancing early warning systems, and (ii) alternative livelihoods: minimizing impacts of climate change for the poor and vulnerable needs to be implemented at the study catchment to help reduce the vulnerability of the people to future climate change and extremes. Other adaptation strategies recommended for the Vea catchment include:

1. Implementation of water use efficiency system as part of effective management of water resources at the catchment. Due to the projected increased in temperature and the variability in rainfall, water use efficiency system such as drip irrigation needs to be promoted to ensure not only all year-round farming but also reduce demands from the three main water sectors: domestic supply, agriculture, and industry.
2. Enhance local capacity to adapt to climate change through improved land use management and agricultural diversification.
3. Awareness creation on the need to protect and effectively manage the Vea dam and streams within the Vea catchment through the creation of vegetation buffers to protect the water resources.
4. Developing early warning systems and farmers access to climate information so as to adequately inform policy makers and enhance farmers' resilience to extreme events as a result of climate change.

Conclusions

The changes in future climate and extremes indices in the Vea catchment was computed from the Expert Team on Climate Change Detection Monitoring Indices based on the observed climate and multi-model mean of four bias-corrected RCMs. The climate simulation under RCP4.5 scenario for the catchment show a warmer climate with 1.3 °C temperature increase in 2020–2049. The trend analysis of the rainfall and temperature extremes indices indicated a continuous extreme temperature rise as well as an increase in intensity of extreme rainfall events in 2020–2049 at the catchment. The projected increase in extreme temperature (TXx and TNx), intensity (R99p), and frequency (CDD) of extreme rainfall indicate the catchment vulnerability to the risk of climate disasters, hence the need for decision-makers to adapt and mitigate climate change over the area. The study also provided evidence that the daily maximum temperature (warmest day) would increase at a faster rate compared to the daily minimum temperature (warmest night), indicating more warming in the day compared to the night. Encouraging adaptation policies such as rainwater harvesting, availability climate information, flood control measures, and the development of early warning systems are necessary at the Vea catchment to mitigate the effects of future extreme climate events.

References

Abeygunawardena P, Vyas Y, Knill P, Foy T, Harrold M, Steele P, Sperling F (2003) Part 1. Poverty and climate change: reducing the vulnerability of the poor through adaptation. World Bank, Washington, DC. Accessed 28 Aug 2019. http://agris.fao.org/agris-search/search.do?recordID=US2012416506

Adeyeri OE, Lawin AE, Laux P, Ishola KA, Ige SO (2019) Analysis of climate extreme indices over the Komadugu-Yobe basin, Lake Chad region: past and future occurrences. Clim Extrem 29:21–39

Alexander LV, Zhang X, Peterson TC, Caesar J, Gleason B, Klein Tank AMG, Vazquez-Aguirre JL (2006) Global observed changes in daily climate extremes of temperature and precipitation. J Geophys Res Atmos 111:D05109. https://doi.org/10.1029/2005JD006290

Amuzu J, Kabo-Bah AT, Jallow BP, Yaffa S (2018) Households' livelihood vulnerability to climate change and climate variability. A case study of the coastal zone, the Gambia. J Environ Earth Sci 8:35–46. https://doi.org/10.13140/RG.2.2.36057.42081

Apata TG (2011) Factors influencing the perception and choice of adaptation measures to climate change among farmers in Nigeria. Evidence from farm households in Southwest Nigeria. Environ Econ 2(4):74–83

Awotwi A, Kumi M, Jansson PE, Yeboah F, Nti IK (2015) Predicting hydrological response to climate change in the White Volta Catchment, West Africa. J Earth Sci Clim Change 6:249. https://doi.org/10.4172/2157-7617.1000249

Bessah E, Raji AO, Taiwo OJ, Agodzo SK, Ololade OO (2018) Variable resolution modeling of near future mean temperature changes in the dry sub-humid region of Ghana. Modeling Earth Systems and Environment 4(3):919–933. https://doi.org/10.1007/s40808-018-0479-0

Bohle HG, Downing TE, Watts MJ (2018) Climate change and social vulnerability toward a sociology and geography of food insecurity. Glob Environ Change 4:37–48. https://doi.org/10.1016/0959-3780(94)90020-5

Chen H, Chong-Yu X, Guo S (2012) Comparison and evaluation of multiple GCMs, statistical downscaling and hydrological models in the study of climate change impacts on runoff. J Hydrol 434–435:36–45

De Pinto A, Demirag U, Haruna A, Koo J, Asamoah M (2012) Climate change, agriculture, and food-crop production in Ghana. International Food Policy 213 Research Institute Policy note #3: Ghana Strategy Support Program, International Food Policy Research Institute. Available: http://www.ifpri.org/sites/default/files/publications/gssppn3.pdf. Accessed 20 Sept 2014

Environmental Protection Agency (2007) Climate change and the Ghanaian economy. In: Policy advice series, vol 1. Environmental Protection Agency of Ghana, Accra

FAO (2013) The State of Food Insecurity in the World 2013. The multiple dimensions of food security. FAO, Rome. ISBN 978-92-5-107916-4

Feng-Wen C, Chen-Wuing L (2012) Estimation of the spatial rainfall distribution using inverse distance weighting (IDW) in the middle of Taiwan. Paddy Water Environ 10:209–222. https://doi.org/10.1007/s10333-012-0319-1

Forkuor G (2014) Agricultural land use mapping in West Africa using multi-sensor satellite imagery. PhD dissertation, Julius-Maximilians-Universität Würzburg

Fosu-Mensah BY, Vlek PLG, MacCarthy DS (2012) Farmers' perception and adaptation to climate change: a case study of Sekyedumase district in Ghana. Environ Dev Sustain 14:495–505

Funk C, Peterson P, Landsfeld M, Pedreros D, Verdin J, Shukla S, ... Michaelsen J (2015) The climate hazards infrared precipitation with stations – a new environmental record for monitoring extremes. Sci Data 2:150066. https://doi.org/10.1038/sdata.2015.66

Gobiet A, Kotlarski S, Beniston M, Heinrich G, Rajczak J, Stoffel M (2014) 21st century climate change in the European Alps – a review. Sci Total Environ 493:1138–1151

Gyasi EA, Karikari O, Kranjac-Berisavljevic G, Vordzogbe V (2006) Climate change vulnerability and adaptation assessment relative to land management in Ghana. Reports of the Members of the Study Team, Accra

Heinzeller D, Olusegun C, Kunstmann H (2016) Projected future (2070–2099): High Resolution (12km) WRF-MPI Daily Near-Surface Relative Humidity over West Africa, (WASCAL Project). https://wascal-dataportal.org/geonetwork/?uuid=0cafe3f0-adf5-43d7-9483-6fbd2b05d86d

Heinzeller D, Dieng D, Smiatek G, Olusegun C, Klein C, Hamann I, Salack S, Kunstmann H (2017) The WASCAL high-resolution regional climate simulation ensemble for West Africa: concept, dissemination, assessment. Earth Syst Sci Data Discuss. https://doi.org/10.5194/essd-2017-93

Kanlisi KS, Arkum TA (2013) Builsa community vulnerability and capacity assessment in the Upper East Region of Ghana. Eur Sci J Ed 9(29):144–164. e-ISSN: 1857–7431

Kim J, Waliser DE, Mattmann CA, Goodale CE, Hart AF, Zimdars PA, Daniel J (2013) Evaluation of the CORDEX-Africa multi-RCM hindcast: systematic model errors. Clim Dyn 42(5–6):1189–1202. https://doi.org/10.1007/s00382-013-1751-7

Kumasi TC, Antwi-Agyei P, Obiri-Danso K (2017) Small-holder farmers' climate change adaptation practices in the Upper East Region of Ghana. Environ Dev Sustain. https://doi.org/10.1007/s10668-017-0062-2

Larbi I, Hountondji FCC, Annor T, Agyare WA, Gathenya JM, Amuzu J (2018) Spatio-temporal trend analysis of rainfall and temperature extremes in the Vea catchment. Climate 6(87):1–17. https://doi.org/10.3390/cli6040087

Larbi I , Hountondji FCC, Mama D, Nyamekye C, Adeyeri OE, Koubodana DH, Odoom PRE, Asare YM (2020) Local Climate change projections and impact on the surface hydrology in the Vea catchment, West Africa. Hydrology Research, In press. https://doi.org/10.2268/nh.2020.022

Lawson E T, Alare RS, Salifu ARZ, Thompson-Hall M (2019). Dealing with climate change in semi-arid Ghana: understanding intersectional perceptions and adaptation strategies of women farmers. GeoJournal. https://doi.org/10.1007/s10708-019-09974-4

Limantol AM, Keith BE, Azabre BA, Lennartz B (2016) Farmers' perception and adaptation practice to climate variability and change: a case study of the Vea catchment in Ghana. Springerplus 5:830. https://doi.org/10.1186/s40064-016-2433-9

M'Po YN, Lawin EA, Yao BK, Oyerinde GT, Attogouinon A, Afouda AA (2017) Decreasing past and mid-century rainfall indices over the Ouémé River Basin, Benin (West Africa). Climate 5:74. https://doi.org/10.3390/cli5030074

Maraun D, Wetterhall F, Ireson AM, Chandler RE, Kendon EJ, Widmann M, ... Thiele-Eich I (2010) Precipitation downscaling under climate change: recent developments to bridge the gap between dynamical models and the end user. Rev Geophys 48:RG3003. https://doi.org/10.1029/2009RG000314

McSweeney C, New M, Lizcano G (2010a) UNDP climate change country profiles: Ghana. Available: http://country-profiles.geog.ox.ac.uk. Accessed 24 Feb 2018

McSweeney C, New M, Lizcano G, Lu X (2010b) The UNDP climate change country profiles. Am Meteorol Soc 91:157–166

Ministry of Environment Science, Technology and Innovation (2013) Ghana national climate change policy. Ministry of Environment Science, Technology and Innovation, Accra

Minville M, Brissette F, Krau S, Leconte R (2009) Adaptation to climate change in the management of a Canadian water-resources system exploited for hydropower. Water Resour Manag 23:2965–2986

Mouhamed L, Traore SB, Alhassane A, Sarr B (2013) Evolution of some observed climate extremes in the West African Sahel. Weather Clim Extrem 1:19–25

New M, Hewitson B, Stephenson DB, Tsiga A, Kruger A, Manhique A, ... Lajoie R (2006) Evidence of trends in daily climate extremes over southern and West Africa. J Geophys Res 111: D14102. https://doi.org/10.1029/2005JD006289

Obuobie E (2008) Estimation of groundwater recharge in the context of future climate change in the White Volta River Basin. PhD dissertation, Rheinische Friedrich Wilhelms Universität, Bonn

Ofori-Sarpong E (2001) Impact of climate change on agriculture and farmers coping strategies in the Upper East Region of Ghana. West Afr J Appl Ecol 2:21–35

Ogountundé PG, Friesen J, van de Giesen N, Savenije HHG (2006) Hydroclimatology of the Volta River Basin in West Africa: trends and variability from 1901 to 2002. Phys Chem Earth 31:1180–1188

Panday PK, Thibeaultb J, Freyc KE (2015) Changing temperature and precipitation extremes in the Hindu Kush-Himalayan region: an analysis of CMIP3 and CMIP5 simulations and projections. Int J Climatol 35:3058–3077. https://doi.org/10.1002/joc.4192

Sarr B (2012) Present and future climate change in the semi-arid region of West Africa. A crucial input for practical adaptation in agriculture. Atmos Sci Lett 13:108–112. https://doi.org/10.1002/asl.368

Seager R, Vecchi GA (2010) Greenhouse warming and the 21st century hydroclimate of southwestern North America. Proc Natl Acad Sci 107(50):21277–21282

Soro GE, Noufé D, Goula Bi TA, Shorohou B (2016) Trend analysis for extreme rainfall at sub-daily and daily timescales in Côte d'Ivoire. Climate 4:37

Sylla MB, Diallo I, Pal JS (2013) West African monsoon in state-of-the-science regional climate models. In: Climate variability-regional and thematic patterns. IntechOpen. https://doi.org/10.5772/55140

Tomassini L, Jacob D (2009) Spatial analysis of trends in extreme precipitation events in high-resolution climate model results and observations for Germany. J Geophys Res 114:D12113

Trenberth KE, Jones PD, Ambenje P, Bojariu R, . . . Rusticucci M (2007) Observations: surface and atmospheric climate change, Chapter 3. In: Solomon S, Qin D, Manning M, Marquis M, Averyt KB, Tignor M, Miller HL, Chen Z (eds) Climate Change 2007: the physical science basis. Contribution of Working Group I to the Fourth Assessment Report of the Intergovernmental Panel on Climate Change. Cambridge University Press, Cambridge, UK/New York, pp 235–336

World Bank (2010) Economics of adaptation to climate change: synthesis report. The International Bank for Reconstruction and Development/World Bank, Washington, DC. Available: https://openknowledge.worldbank.org/bitstream/handle/10986/12750/702670ESW0P10800EACCSynthesisReport.pdf?sequence=1. Accessed 26 Feb 2018

Xu CY (1999) From GCMs to river flow: a review of downscaling methods and hydrologic modeling approaches. Progr Phys Geogr 23(2), 229–249. 107, 21277–21282. https://doi.org/10.1073/pnas.0910856107

Zhang S, Lu XX, Higgitt DL, Chen CTA, Han J (2008) Recent changes of water discharge and sediment load in the Zhujiang (Pearl River) Basin, China. Glob Planet Chang 60:365–380

6

Just Societal Transformation: Perspectives of Pastoralists in the Lower Omo Valley in Ethiopia

Sabine Troeger

Contents

Abstract

Pastoralists' livelihoods in Africa are highly endangered by adverse forces – the climate change being one among those. Against this background, climate change adaptation is conceptualized as strategic agency in the field of risk-laden livelihood environments, that is, agency in the face of risky options and non-calculable uncertainties.

The chapter conceptualizes pastoralists' livelihoods exposed to a four-fold hierarchy of environmental risks and forces defining the actors' arena of strategic decision making: From the global scale of ever extending impacts by the climate change imperative, to the national scale of government policies in terms of decentralization, challenging people to govern and define their communal

S. Troeger (✉)
Department for Development Research, Geography Institute, University of Bonn, Bonn, Germany
e-mail: troeger@geographie.uni-bonn.de

efforts in terms of climate change adaptation, and down to the regional scale, which in the presented case is dominated by a large-scale investment, the Kuraz Sugar Development Project, which again confronts local actors with adverse forces toward villagization and eviction from pasture grounds. Right at the end of this hierarchy and in accordance with discourses on "climate services," the end-users and local actors, the pastoralists, are confronted with and offered a product that they can input into their decision making: cattle feed from the residues of the irrigated sugar cane. The question remains whether substantive aspects of processes turning into true environmental and social justice in terms of recognition, procedures, and distribution will be paid attention to.

Keywords

Societal transformation · Climate services · Environmental justice · Riskscapes · Decentralization · Land grabbing

Transformation Toward Sustainability in Terms of Social Justice: Two Perspectives in Ultimate Dependence

The notion of transformation is increasingly promoted in scientific and likewise popular discourses as a solution to unsustainable practices. Transformations toward sustainability are needed to address many of earth's profound environmental and social challenges. In transformational adaptation literature the distinction between incremental adjustments and transformational responses is to be found. As defined by the IPCC-Report 2018: "Global Warming of 1.5 °C," incremental transformation refers to some means and instruments of "adaptation that maintains the essence and integrity of a system or process at a given scale," while "transformational adaptation implies 'changes of fundamental attributes of a socio-ecological system." A transformative response to obviously unsustainable circumstances of life, be these ecologically or socially based, will require engaging with the root causes of inequality and likewise environmental degradation, which necessitates some transformational adaptation in the indicated sense. Furthermore, actions taken to shift socio-ecological systems toward more sustainable pathways can have substantial social impacts and may exclude members of communities from decisive processes.

The named perspective emphasizes the importance of contesting dominant social and political structures, rather than accommodating change to address the root causes of unsustainable systems and catalyze genuinely alternative futures. Transformations toward sustainability can only be taken as a success in case social justice is considered a central concern. In consequence, all actions taken to shift toward environmental sustainability can have both positive and negative social implications for different groups relative to the status quo, which means attention is needed to both understanding and realizing social justice during sustainability transformation (Temper et al. 2018). Following this notion three dimensions of justice are to be differentiated: (a) recognitional justice in acknowledgment of and respect for preexisting governance arrangements as well as the distinct rights, worldviews, knowledge, needs, livelihoods, histories, and cultures of different groups in decisions; (b) procedural justice with reference to the level of participation and inclusiveness of decision making and the quality of governance processes, and c) distributional justice, which can be defined as fairness in the distribution of benefits and harms of decisions and actions to different groups across space and time (Bennett et al. 2018).

Following the vision to turn scientific information from climate monitoring, research and modeling into operationally available information and services that would help societies to cope with climate variability and change, the World Meteorological Organization (WMO) launched the process for developing the Global Framework for Climate Services (GFCS) at the World Climate Conference (WCC) in September 2009. This conference recognized ongoing discourses on the unambiguous importance and relevance of climate services, which were considered essential to adaptation to environmental and climate variability and change (European Commission 2014). Climate services are to be seen as one essential contribution and element of transformations toward sustainability. But at the same time the perspective of climate justice identifies climate services as the source of a double inequality with a distribution of risk and responsibility (Adger et al. 2006, 2009). Climate services will mitigate the impacts of climate-related stresses, while the end-users' perspective is a key in tailoring the service measures (climate-services.org).

Against the background of these generalizing recognitions with relation to societal transformation in terms of environmental justice the chapter aims to explore the implications of present day environmental dynamics, the climate change intertwined with market and governance perspectives exemplified in their concretized meaning for pastoralism and pastoralist livelihoods in the Lower Omo Valley right in the south of Ethiopia: The Lower Omo Valley, until 2011 one of the most peripheral and likewise politically neglected regions of Ethiopia, has turned into a hotspot of societal transformation geared by natural, social, economic, and cultural forces. Home to 16 ethnic agro-pastoralist societies (recognized UNESCO World Cultural Heritage in 1980), formerly well adapted to the fragile semiarid environment of the lowlands, the Lower Omo Valley is nowadays highly impacted by irrevocable and fundamental changes in livelihood constellations caused by forces in a four-fold global to local scale gearing toward some ultimate and irrevocable societal transformation, that is, processes, which hold a strong grip on those ethnicities calling the Lower Omo Valley their home in terms of fundamentally re-defining the constituents of livelihood systems as of at present. With reference to the initial perspective of the chapter the argument will pose the question in how far and to which degree processes of transformation to be observed in the Lower Omo Valley are to be taken as sustainable, which includes the perspective of justice with reference to the addressed ethnicities, the pastoral communities in their distinct and unique representation.

Grounded on actor oriented and constructivist reasoning the chapter subsequently relates to the perspective of human agency in the face of risky environments. Following Appadurai's concept of spatiality (1990), when he coins five terms with the suffix "-scape" indicating the ambiguity and fluidity of social phenomena, the chapter interprets agency by people in the context of their "social constructions" of their environments. The argument relates to the idea of "riskscapes" as defined by Mueller-Mahn and Everts (2018). Riskscapes are intertwined with social practices. The focus on practices posits that understanding human activity is key to any deeper knowledge of events and states. Schatzki (2002) emphasizes the fact that any practice involves various people and is always part of a larger set of socially ordered actions, highlighting the dynamic nature of social practices, how they change and how they relate to spatial and temporal dynamics. The argument follows Schatzki (2012:14), when he says, "a practice (. . .) is an open ended, specially temporally dispersed nexus of doings and sayings." Based on this recognition the concept of "riskscapes" refers to temporal-spatial phenomena that relate risk, space, and practice. Riskscapes link the material dimension of physical threats, like for example the climate change, the discursive dimension of how people perceive, communicate and envision risks, and the dimension of agency, that is, how people produce risks and manage to live with them. In consequence, risks are not to be taken "objectively given" phenomena, but are shaped and constantly modified through practices, which are socially embedded. Thus, observations in the field and with representatives of the communities in focus are to be understood in terms of agency in recognition of people's perceptions, interpretations, and evaluations of their environments.

The chapter elaborates on four perspectives of given environmental – natural as well as social and sociopolitical – constellations and livelihood frameworks in scales in the Lower Omo Valley. The argument will draw the attention to two comprehensive conclusions of these observations and assessments in the field, to subsequently link these with regional forces in terms of villagization and sedentarization policies, which are taken as "dispositives" as understood by Foucault (1977). In line with this understanding, present and pressing development perspectives are outlined, which highly challenge and tackle the field of environmental justice. Concluding, the argument will revise the field and discursive findings with relation to the four analytical scales and suggest some rounding up picture of the observed societal processes in terms of a transformation answering the claim to social justice.

Figure I titled "Societal Transformation" illustrates the above referred to analytical concept in scales:

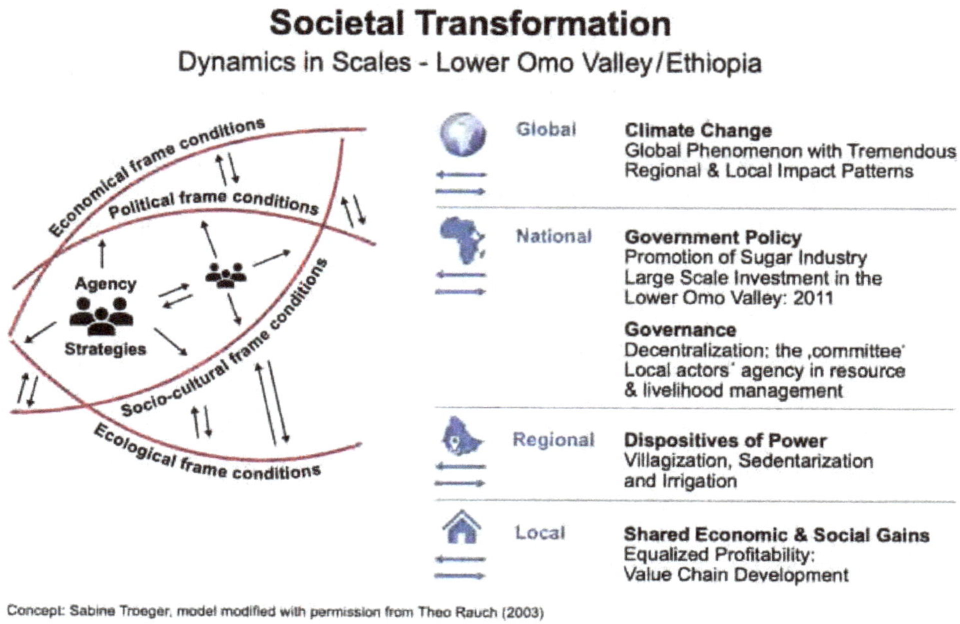

Figure I: Societal Transformation, Dynamics in Scales – Lower Omo Valley/Ethiopia (source: Concept Sabine Troeger, model modified with permission by Theo Rauch – 2003)

The Climate Change Imperative as Perceived and Acted upon by Pastoral Communities: The Global Scale

Data from the field leave no room for doubt about the severity of ongoing and in the course of time further enforced climate change impacts in the Ethiopian Lowlands (Troeger et al. 2012). The Lower Omo Valley, located in the southern part of these lowlands, is highly exposed to a climate change pattern characterized by considerably shortened *Belg* rains as part of the customary rainfall pattern. These were formerly expected to be starting end of February or early March and lasting until end of May or beginning of June. The eastern and southern regions of Ethiopia, demarcated by the Rift Valley, are highly dependent on this specific rainfall pattern for their annual agricultural production as well as their migratory routes with their livestock, which in this case is cattle, whereas the goats and sheep, the smaller ruminants, remain with the women, the wives, children, and elderly people. They will feed on the pastures and bush-vegetation in the vicinity of the homesteads, which are rhythmically shifted within a radius of some kilometers but remain in line with the pastoralist livelihood system more or less concentrated on some an ethnically defined neighborhood.

At the time of the field assessment the pastoralists discussed the spatial and quantitative changes of temperature, precipitation and floods. To their perception, all of these un-contradictorily stated deviations and irregularities in rainfall started around the year 1989 and since then have shown a gradual decline in precipitation. More severe changes were perceived to have started after 1998 and culminated for the first time in the year 2011, when large parts of the Horn of Africa were struck by one of the worst droughts ever recorded in 60 years, which translated into a severe food crisis. In July 2011, the United Nations declared a famine, but already as early as January 2011, considerable heads of livestock had died, as reported by the people. The drought that started in 2011, whether it was caused by global climate change factors or the effect of a very harsh La Niña event as suggested by IRIN News (15 July 2011), took place in the wake of 4–5 years of more or less failed *Belg* rains, as summarized to be a general trend by the diagram below:

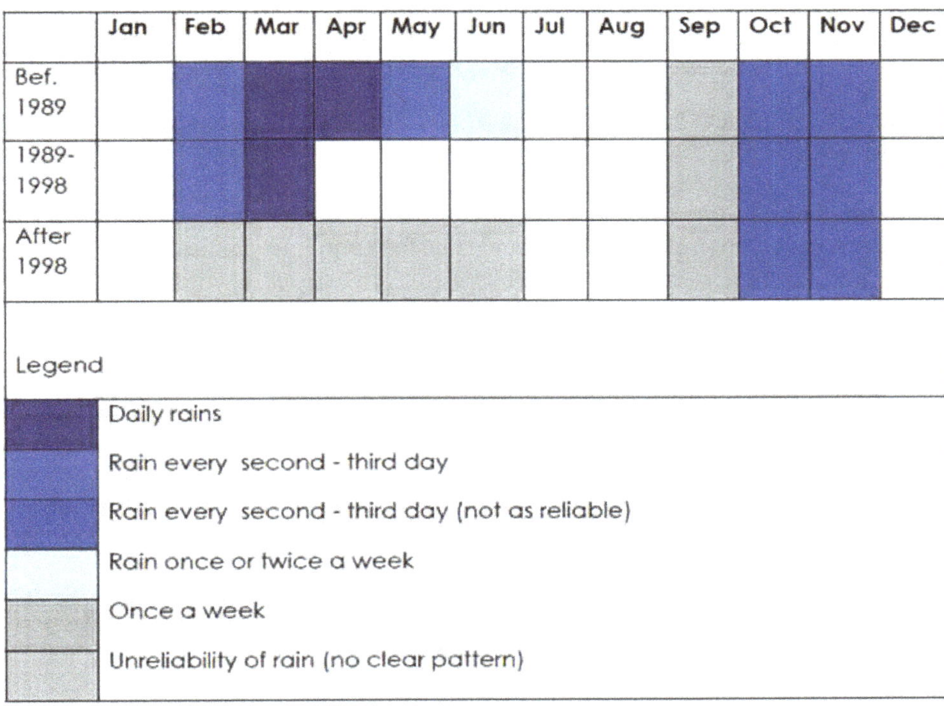

Figure II: The Climate Change pattern as perceived and acted upon by the people. (Source: Troeger et al. 2012)

"*Everything that is happening now, is beyond our capacity*" was the unisonous statement of the people, be these young or old, when talking to them in the field. The conclusion drawn by Moro, at that time Nyangatom women leader, summarizes the research findings (Troeger et al. 2012) and unambiguously indicates a livelihood configuration highly at risk, going hand in hand with a dissolution of beforehand sustainable codes of conduct, namely the indigenously established "culture of sharing," when people deliberately and spontaneously shared food and further agricultural produce and items (Troeger 2016b & 2018).

A concordant climate change pattern showed the precipitation and subsequent constellations of food crises in the past year, the year 2019, which can be taken as an indication of the further continuing decline in and unreliability of rainfall:

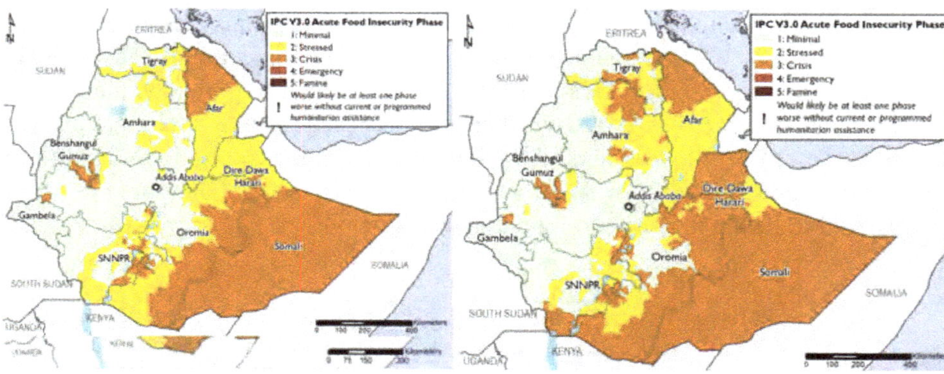

Figure III: Patterns of food Insecurity in the wake of deficit precipitation in the Lowlands – the year 2019. (Source: with permission by Famine Early Warning Systems Network (FEWS.net))

Moving back to the initial question of people and pastoralists in confrontation with processes of environmental risk and transformation, pastoralists in the Lower Omo Valley see themselves and are to be taken as threatened beyond their formerly established coping potential. According to their reports the young herders could not follow their seasonal migratory pattern among other things. There were reports about young herders, who stayed away from their culturally manifested neighborhood for several seasons in succession. People were worried about these environmental tendencies and complained that, as the weather got drier and it became increasingly difficult to survive, neighborhoods fell apart, people abandoned their traditions and were forced to accept other lifestyles (Troeger et al. 2012).

The "Committee": A Pathway Toward Climate Change Adaptation and Mitigation? – The National Scale

Turning to the national level, the argument moves on from the following observation:

> Following the agenda of the government, the claim of "development" has always been and is as well with the new policies guided by the present Prime Minister Abiy Ahmed the directive of the rulers of the state. Governments' interventions labeled NRM, natural resource management, combine hegemony and governmentality projects to conform citizens into its governmental state ideology. While local governments continue to be representatives of the controlling apparatus of the political elite rather than representatives of interests of their constituents (Ayele 2011), they will follow the "development narrative." Party loyalty is demonstrated by adopting agricultural ideas promoted by the state, which is indicated by Lefort (2012), who suggests that so-called model farmers are rather party agents instead of famers with high agricultural competence.

Based on this insight the political role of "committees," coinciding with the claim for decentralization all over Africa and together with the mainstreaming of participatory approaches in development practice, is to be questioned. Do communities experience a shift in policies to advocate local actors and resource users and play a more active role in natural resource management (NRM), which in many cases today is to be understood as climate change adaptation (CCA)? Decentralization in general describes the process by which bundles of entrustments, like regulatory and executive powers, responsibility and authority in decision-making, are transferred to local agents, which again calls for new institutions and processes of institution building, that is, the committees. The following argument digs deeper into this perspective of local policies and into the perspective of "handing over the stick," the guiding principle of the participatory approach in development practice PRA and of governance in local contexts. And at this point, the question of environmental justice becomes vital as the impact of decentralization is to be measured with respect to recognitional, procedural, and distributional just and fair outcomes relating to actors and activities on the local scale.

The targeted processes of decentralization come about in the shape of "committees," that is, institutional settings, which have mushroomed up in various development contexts all over Africa. Activities of the "Western" ideal of democracy in the shape of committees focus in general environments of obvious global change, which force local actors to unanimously and fundamentally adapt to the irreversible changes in nature. Informed by this perspective, the committees mirror the principal idea of equal representation and equal voice. Per definition has the ideal of "participation" developed to be the counter-balance to the sociopolitical institution of "committees," which again claims to "give people a voice." In consequence, the notion of "voice" unambiguously links the concepts of "committee" and "participation" with the need for "communication," true communication, which enables intersubjectivity via a mutual exchange of people's perceptions, interpretations, and valuations, in short: social constructions. In summary, committees are taken to be the platform for participation in realization of the demand for "communication" regarding "giving people a voice."

Corresponding with Government policies in promotion of sustainable and decentralized community development in Ethiopia, a "committee" has been established with the Nyangatom community. The Nyangatom are one of the 16 ethnicities in regional focus. Their committee was explicitly established to safeguard and control the recently established scheme of rangeland enclosures in support of climate change adaptation needs in terms of referred to propagated improved land use techniques. Rangeland enclosures, enclosed by bush thorn fences, are manifestations of private landownership, which excludes members of the community from participating in the advantages of the protected grazing areas behind the fences. In consequence, parts of the former "commons" are claimed by the powerful representatives of the communities. The committee was first installed in order to govern the new scheme on local grounds and resist against the powerful take over by some community members. But very soon after it was appropriated by Government employees, by "experts" from Addis Ababa.

In contradiction to the outlined and to our "Western European" eyes normalized and customized ideal of democracy the established committee meets societal constellations on the ground, which had known and were based on some means and ways of mutual consensus finding by their own constitution. The Nyangatom, one of the unique pastoralist ethnic groups of Ethiopia's south, situated in the Lower Omo Valley, numbering about 10,000 people, has like its neighboring pastoralist communities, ever since been guided by an "age-group system." The system is constituted by five age groups: children, youths, herders of 25–50 years, the political class of the "elephants" aged 50–65 years, and the very old and wise men in the role of spiritual guides, as explained by two community guides:

> The elders used to manage the grazing land by telling the herders to shift from pasture to pasture. They sent young herders to check the condition of the grass, and then they decided on the migration paths. If someone was seen taking his animals to the forbidden areas, he would be punished. But punishment is not constant. It will vary from people to people and it depends on the severity of the action. But it as well respects the livelihood constellation of the trespasser and defines the punishment in accordance. (Soya Kurupa, man – 14 March 2015, approx. 42 years) (Troeger 2018)

> The cultural management has no written rules and regulations, because once the elders pass the decision, no one will disobey. It is a management, where everyone has to obey. We have grown up with it, so we like it very much. (Moru Lomarle, woman – 12 March 2015, approx. 55 years) (Troeger 2018)

Interpreting village-level ethnographic data from the named pastoralist livelihood scenery, the case of the Nyangatom, the chapter supports the following thesis: The formula of "transformative adaptation," as propagated by, for example, the 5th IPCC assessment report 2014 and likewise by the report on "Global Warming of 1.5 °C" (see above) with some successful climate change adaptation in view, and subsequently multifold suggested by initiatives hailing processes of societal transformation, does not necessarily gear toward a change in the name of a more resilient and sustainable development. Field data reflect on interpretations and perceptions by the pastoralists involved relating to this transformative impetus. They indicate "disagreement in the shadows" as defined by Rancière (1998), and a disruption of processes of communication and mutual cooperation. With this portrayal Rancière indicates some community constellations, which do not allow for open resistance but only offer some hidden away and not openly articulated opposition (ibid.).

Listening to the people, to "the people's voice," reveals the true process lines of the committee meetings and offers the chance to get closer to the meaning of "governance" and "participation" facilitated by processes of communication in given frameworks of the committee. Perceptions and interpretations are singled out in line with the perspectives of: "structural forces and governance," "modes of participation," and "articulation of social correctness."

> I myself had the idea of establishing rules to govern our problems with the drought and the use of the pastures, but then I heard that the Woreda people are coming to establish the rules and regulations. And there were some people from Addis Ababa, who gave us some papers,

> which listed the rules. They asked the people to add on it, and of course, everything was clear, and nobody added anything on top of the by-laws, and the committee was selected from different age-groups. (Lobowa, approx. 42 years – 14 March 2015) (Troeger 2016b & 2018)

Quite in line with the general articulation of the "development narrative" in Ethiopia, obviously no participation, which could indicate true "governance," had been sought. The community member claims initially own, self-determined ideas about necessary regulations, but he had no chance to promote these tentative frameworks of governance. No "freedom to articulate one's own voice" is given in this process of communication. No recognitional nor procedural justice is given to him and his ideas. He is neither respected in his worldviews, knowledge, and needs, nor is there any true chance for true participation offered to him. The message from the district and national governance representatives turns out to be an unambiguous top-down message, which has to be "swallowed" by the people on the ground, by those, who in future would be the ones to realize these handed over regulations. Structural forces govern the information instead of true communication processes. No communication in the ideal sense is realized, which is even more emphasized by the following statements:

> Us elders, we were unsure whether to accept the new rules, but the Woreda people made us accept the idea of bye-laws. (Dida Lolibes, man – 11 March 2015, approx. 64 years) (Troeger 2018)

> The leaders from the Woreda and the people from Addis made us discuss the question of the punishment for not following the rules. So, the discussions were led by the experts, and they were the ones to give a speech on the positive side of the new governance system and make the community accept the idea. Even me, I accepted it. (Ariapa Lokitibo, man – 14 March 2015, approx. 38 years) (Troeger 2018)

By listening to the people's voice it becomes obvious that the modern idea of committees will not match with the culturally established and believed in articulation of authority and representation. The idea puts the indigenous identity at risk. The articulated obedience toward culturally manifested rules represents customary regulations and a sense of belonging and identity. Especially the emphasis on a fair, socially reflected and balanced execution of established rules and regulations contrasts the perception of today's realization of power and authority. The obvious notion of "disagreement" indicates a situation, when meant to be partners in communication do not necessarily contradict each other openly but disagree in the shadow of claimed consensual modes of governance. On the surface, this consensual governance has reduced political conflict and disagreement (Swyngedouw 2011), but in reality, the partners in communication have not reached a common ground of understanding. "Disagreement in the shadow" should be taken to the open and made understandable to all. The new challenges and arrangements with relation to the range enclosures and in consequence processes of commodification of the 'commons' are to be discussed and reflected on mutually and in an atmosphere of mutual freedom of articulation. But this disposition cannot and will not be found in a situation, when the inherited and traditional authority of the "elephant" age group is confronted with some authority of the government in regional and national contexts.

The so-called "participatory" processes in the pastoralist communities does not only confront the people on the ground with readymade answers and new regulatory frameworks but excludes some members of the community completely from the wished for democratic process. They are put at risk. Thus, neither some "equal grounds of understanding" nor any "freedom to articulate one's perceptions" and evaluations of the given social needs in terms of a fair and transparent handling of the new "structural forces" in the shape of the climate change imperative is sought for:

> My neighbors in our settlement did not participate in the discussions. According to what I now know they are not happy with the decisions because they did not get any clarification on this matter like we did from the Woreda and the people from Addis. (Ariapa Lokitibo, man – 14 March 2015, approx. 38 years)

The example of the Nyangatom pastoralists in south Ethiopia unambiguously highlights given processes of communication to be by far not manifested in the open and freed from structural power. Communication in the committees is dominated by "experts," that is, the agents of the government following the "development narrative" as outlined above. The actors on the ground resisted the officially outlined principles and practices of climate change adaptation "in the shadow": "We should take disagreement to mean a determined kind of speech situation: one in which one of the interlocutors at once understands and does not understand what the other is saying. Disagreement is not the conflict between one who says white and the other says black. It is the conflict between one who says white and another who also says white but does not understand the same thing by it or does not understand that the other is saying the same thing in the name of whiteness" (Rancière 1998).

Essentially, this argument proposes that governing climate change has been characterized not only by disagreement, but climate change communication in this case is to be understood as subcutaneous resistance, voiced quietly and under cover, stating in the end some ultimate failure in communication. With reference to the differences in an articulation of social justice as defined above the committees in today's shape are not to be understood as windows of opportunity toward more environmental justice and sustainability in societal transformation. Mouffe (2013) links to the reflections of Rancière (1998) and Swyngedouw (2011) referred to above, when she states: "We have to relinquish the claim that the processes of democratization should consist in the global implementation of the Western liberal democratic model . . . the kind of individualism dominant in Western societies is alien to many other cultures, whose traditions are informed by different values. Democracy, understood as 'rule by the people', can . . . take other forms – for instance, forms in which the value of community is more meaningful than the idea of individual liberty."

The argument summarizes observations in the field accompanied and highlighted by the voices of the local agents stated above and objects to any claim of social justice granted to these agents, the pastoralists. The process of decentralization is not to be taken as any granted path toward true communication with effective recognition of all community members on equalized grounds. Accordingly, the challenge to reach some just and in this way sustainable societal transformation is not responded to in any meaningful terms of governance.

Large-Scale Investment, Irrigation, and Villagization: Opening the "Ground" to the State – The Regional Scale

Turning toward the regional scale, the argument links to the frequently quoted statement by the late Prime Minister of Ethiopia, Meles Zenawi, the sociopolitical meaning of which is to be considered one of the most crucial turning points toward the situation of societal transformation as of today: "I promise you that, even though this area is known as backward in terms of civilization, it will become an example of rapid development" (Meles Zenawi speaking at the Pastoral Day Festival in Jinka, South Omo Zone, 25 January 2011).

The first "state" representation in the Lower Omo Valley times back to the 1960s, when programs of sedentarization and the establishment of the Omo National Park (1966) and the Mago National Park (1979) led to increased resource pressure for the agro-pastoralist ethnic groups, traditionally inhabiting this area and using its resources for pastures for their livestock and especially the river banks for river retreat agriculture. Soon after, that is, in the 1980s, the then Ethiopian military regime released the Omo Basin Development Plan and in this context initiated the construction of the Gibe III hydro dam north of and at the entrance of the Lower Omo Valley, the construction of which was started in 2008 and the filling of which was completed in 2016. Today, the Gibe III dam is to be taken as the most influential structural incentive of those transformative forces and risks, which have defined and continue to define life and the livelihood systems in the Lower Omo Valley.

Relating to Meles Zenawi's argument, commercial investment processes were predicted and initiated, which turned Ethiopia and especially the pastoralist lowlands of the country into a hotspot of what commonly is called the "land grab." By emptying Ethiopia's pastoral lowlands for large-scale investors, in the case of the Lower Omo Valley the Kuraz Sugar Development Project (KSDP), the ground was opened for police and military, for the "state," giving protection to these investors. The establishment of the large-scale investment was sidelined and backed by the implementation of new agricultural technologies in settling and sedentarizing pastoralists in relocation sites by providing them with agricultural extension packages in terms of chemical fertilizer schemes, improved seeds, which the pastoralists had to pay for, and small parcels of irrigated land (Regessa et al. 2019). The government deliberately claimed the culturally inherited pastoralist land, the "commons," for investments, as it was considered "underutilized" and "unproductive," as the pastoralists were interpretated not "civilized" enough to make appropriate use of it. This villagization program, officially known as *Mender Masebaseb*, meant some fundamental transformation to the pastoralists – they were supposed to become and be sedentary and give up or at least considerably reduce the basis of their livelihood system, their seasonal extensive cattle drives and migrations.

The Lower Omo Valley, South Omo, in historical times was and up to now is to be taken as an environment of latent and open social conflicts, conflicts on various levels and with reference to diverse agents and contesting parties. The 16 agro-pastoralist ethnic groups find themselves in partly violent competition on grazing grounds and heads of cattle. These conflicts have largely been analyzed by anthropologists like David Turton (1985) and Jon Abbink et al. (2014). Nowadays, since 2011, since the valley has turned into a region of high national interest, the local manifestation of the Government-owned Kuraz Sugar Development Project has opened new grounds for conflicts including human rights violations. About 90,000 people in the Lower Omo valley had been depending on flood-retreat farming for sorghum cultivation along the Omo (Turton 2010). In the more southerly parts of the Lower Omo the areas flooded by the river were larger, while the people on the surrounding plains were even more reliant on flood retreat farming than their neighbors in the north, as these practice rain-fed agriculture. Since the end of the Omo floods in 2015, most groups in the northern part of the Lower Omo Valley have become entirely dependent on rain-fed cultivation for crop-production, which leaves them highly vulnerable to climate change induced droughts, or solely dependent on corporation-controlled irrigation systems (Stevenson and Buffavand 2018). Furthermore, reductions in the river flows downstream of the sugarcane plantations are taken to reduce the availability of water – moisture in the soil – with damaging effects on the pastures for the pastoralists' livestock, especially cattle and sheep, while goats are browsers and can partly feed on bush vegetation. Land clearance for sugar plantations has additionally restricted wildlife in vital habitats and cut people's access to subsistence opportunities (Buffavand 2016).

KSDP operates on about 100,000 ha production sites of sugar cane, the formerly pastoralist "commons," irrigated by water from the Gibe III hydrodam. In conjunction with the dam, a system of canals has been built to distribute water to the Omo-Kuraz Plantation, within which selected plots of land have been assigned for use by "settler pastoralists" (Pictures I).

Picture I The example of one of the two irrigation canals along both sides of River Omo (source: with permission by Omo Turkana Research Network (OTuRN 2016)

At full capacity the Kuraz production sites would realize, this the initial projections, a quarter of Ethiopia's sugar and ethanol production, more than any other sugar production site in the country (ESC 2014). The sugar cane is, or will in near future be processed in up to five sugar factories lined along river Omo and situated in pastoralist territories. What is happening in the Lower Omo Valley combines elements of different types of displacement. The scheme was planned and executed by the current regime in Ethiopia in collaboration with an Italian engineering firm and international financiers. In 2012, the South Omo Zone Agriculture Bureau promoted a "Villagization Plan," which described how and when in the course of the coming year the people would be moved into new villages (FDRE 2012). In the plan the benefits of villagization are described as follows: "growing new crops, such as sugarcane on irrigated fields would improve food security; providing people with schooling and medical services would improve their health; settling in planned villages would reduce conflicts that result from "mobility in search of water and pasture". . . . "In the case of land that relies on rain (for cultivation) 2 hectares would be appointed to each settler pastoralist, whereas those settling on irrigable land would be appointed 0.5 hectares each" (FDRE 2012).

In contradiction to these positively connoted outlooks, findings from the field by E. G. J. Stevenson and L. Buffavand (2018) indicate that crop yields were and will be by far not sufficient to feed the pastoralist settlers, let alone to produce a surplus for sale. ". . .villagization was actually experienced by the Bodi settlers as undermining food security and diminishing well-being. It was experienced as heat, as disruption of routines, as a loss of control over livelihoods, as being treated like children rather than adults, and as abstracting work from ordinary sociality" (ibid). In congruence with these findings the study on "Social-ecological change in the Omo-Turkana basin: A synthesis of current developments" by almost the entire group of researchers active in the Omo-Turkana basin (Hobold et al. 2019) recapitulates their findings in a sense of comparative winners – the migrant laborers – and losers – the indigenous peoples (including both agro-pastoralists and pastoralists, but with more borne by agro-pastoralists currently): "Important SES-wide impacts will result from changes in regulating, provisioning, and cultural ecosystem services, including potential environmental degradation; loss of biological and cultural diversity; heightened competition and conflict over natural resources; and the potential for increased dependence on food aid" (ibid.). The other major actors – agro-industry investors, and the governments of Ethiopia and Kenya – they classify as in an equivocal position, with potential for large gains, but also exposure to substantial risks – a position, which is highly contradicted by one of the authors – Benedikt Kamski – at another place. In his argument titled "Omo investors won't scrub away Kuraz's sugary stain" (Kamski 2019) he states: "ESC (Ethiopian Sugar Corporation) rushed into the multi-billion dollar project, largely neglecting critical feasibility and impact assessments" (ibid). The author stresses the urge ESC is experiencing when the capital-intensive transformation of the region failed to bear fruit for its indigenous communities, which in turn creates new conflict dynamics. He concludes: "If economic interests alone guide investment decisions, official visions for inclusive pastoralist development – as recently reiterated by Deputy Prime Minister Demeke

Mekonnen – will require additional effort and financial commitment by the government, and a big question mark remains over whether this will materialize" (ibid.).

Climate Services, Pathways Toward Processes of Beneficial Societal Transformation? – The Local Scale

As indicated above, the chapter will now and finally turn toward and highlight some projected initiative in terms of climate services towards a promotion of livelihood security and environmental justice in favor of the indigenous inhabitants of the Lower Omo Valley, the pastoralist societies. The reasoning of this perspective responds to the again and additionally outlined and named below livelihood risk and conflict scenarios taken to be highly virulent and tangible.

Upon the time and again scheduled and rescheduled completion of the Kuraz Sugar Development Project (KSDP) the overall about 800,000 ha pastoral land will be reduced by 100,000 ha. The sugar production was from the very beginning meant as vehicle of state governed development and progress. But, in spite of the major cutbacks in size and processing capacity, has the KSDP highly impacted on the livelihood system of the pastoralists (Kamski 2019; The Oakland Institute 2013). River Omo is nowadays framed by about 15-m broad irrigation canals, which hinder traditional use of water by the pastoralists for animals and river bench agricultural cultivation. As the pastoralists had to give way to the sugar plantations, they were called to re-settle in newly established permanent villages – the "villagization program" driven by Government policies. In the course of the thus induced transformation processes, many pastoral households are nowadays facing a livelihood situation of food-insecurity. In need of watering their livestock the pastoralist herders are nowadays forced to virtually beat their animals through a dark subway underneath the bottom of the irrigation canals, the so-called underfly. As the floor of these subways is moody and slippery, herders complain about injured cattle with broken legs. Lacking mostly veterinary services in the pastoral setting of life herders will eventually feel forced to kill their injured animal (Pictures I and II).

Additionally, and along with the establishment of the sugar plantations some extensive labor force was needed. Up to today, about 30,000 migrant laborers have arrived from the Ethiopian Highlands. Moreover, re-settlers from the highly populated areas of, for example, the Konso in South Ethiopia have taken advantage of the labor demand in the sugar project. The number of migrant laborers is expected to reach some hundreds of thousands, when all production sites and the five sugar factories will be fully operational (Kamski 2016).

Against the depicted situational setting, resource-based conflicts and at times violent encounters are and in future will be the "order of the day." In the course of resource – land and water – scarcity, the agro-pastoralists ethnicities find themselves squeezed by natural factors (climate change) and the sugar cane development project. Hence, some growing resentment and animosity toward the dominance of the sugar project and some resistance against the migrant population are poisoning today's encounters between the diverse interest groups. The pastoralists feel treated

Picture II One of the 'underflies', the only paths for the pastoralist herders to drive their livestock for watering (source: with permission by Omo Turkana Research Network (OTuRN 2016)

unjustly and surely suffer from a lack of recognitional as well as distributional justice.

Moreover, resource scarcity has increased the intensity and frequency of interethnic conflicts among the agro-pastoral ethnicities. If not solved, the current tensions and violent confrontations are likely to engulf the entire South Omo regional entity – one of Government's perspective development corridors in Ethiopia.

In summary, food and livelihood security of the agro-pastoralists in the Lower Omo Valley have become highly vulnerable. This relates on the one hand to the considerable impact of climate change in the course of which the *Belg* Rains fail several seasons in succession as outlined above but likewise to the reduction of freely accessible pasture land in the course of the establishment of the sugar production sites and thus the reduced option in shifting cultivation in comparison to former times. The anchor element of this setting is "feed for the cattle," the central economic and likewise sociocultural value in pastoralist livelihoods. With relation to this socioeconomic "anchor element" in question, the agricultural value chain of the produce sugar cane moves into focus. This value chain is centered to be the protagonist of conflict mediation and resolution in its function to smoothen livelihood stress and vulnerabilities. It is meant to be the manifestation of climate services applicable in given contexts.

In the course of the following argument some design shaping a rather realistic feature of climate services was outlined in 2019, but seemingly has not been realized up to now. As pointed out above the following discussion will in consequence be somehow speculative – but all the same realistic. The envisioned scheme of climate services targets the given lay of conflicts between the indigenous pastoralist communities and the Ethiopian Sugar Corporation (ESC) by one central element: by cut-off residues, bagasse, and molasses from the sugar cane, by-products from the sugar production – to be turned into cattle feed. A project like this would be conceptualized as "contract research," with high emphasis on the value chain development of the sugar cane, but likewise with explicit focus on various social perspectives on conflict resolution, discharging into a process of integrative societal transformation (Picture III).

Guided by the promoted Government politics in terms of "ethnic reconciliation" the project could operate on value chain development and open access to cattle feed for the pastoralist ethnicities. New environments of coexistence among the interest groups in question on equalized grounds were to be defined and processes of integrative societal transformation to be eased.

It can be foreseen that climate services as outlined above bear dangers in terms of some processes of fatal societal transformation, as will be outlined as follows: Indigenous pastoralist societal organization does by its concepts of live and livelihood constituents not incorporate the idea of feeding and, even much less, fattening cattle, whereas the government program of villagization claims the idea of sedentism of the pastoralists in South Omo, which necessarily implies the need for some ways and means of stationary sources of income for the then non-migratory population. In consequence enforced societal transformation will demand considerable conceptional architecture and research endeavor in order to accompany and facilitate these pathways into newly to be designed livelihood systems and outcomes. But, this

Picture III A herder with his favorite ox (source: with permission by Julia Pfitzner)

should be noted: these at first sight completely new and seemingly culturally non-acceptable pathways are not to be taken as any purely unrealistic illusion! Already ten years ago and in the face of the above illustrated severe climate change induced droughts, had Nyangatom cattle herders started to sell cattle – at that time under disguise and hidden away from their fathers and elders (Troeger et al. 2012). Very obviously, processes of societal transformation were on the way – already at the beginning of the decade, to be enforced and accelerated by today's bundle of urges. Already at that time some members of the pastoral communities, especially the young men and women, were ready for processes of societal transformation. They were for example ready to abandon their culturally manifested and strict rule to not slaughter cattle except for religious reasons. The perspective of cattle feed would newly and additionally demand some readiness for societal transformation on the side of the pastoralists. The pastoralist cultural setting does by traditionally fixed rules not allow for any sale and ownership in capitalist terms, which would necessarily be one of the terms of sale and storage of cattle feed. – And at this point the question whether the climate service initiative in the outlined way would do good or harm must stay unanswered.

Conclusion

Concluding and turning back to the initially highlighted challenge envisioning societal transformation in terms of environmental justice realized with and for the inhabitants of the Lower Omo Valley in Ethiopia the chapter summarizes the argument in succession of the four environmental scales of analysis.

Very obviously do the forces on the global scale, the climate change imperative, threaten the livelihood system of pastoralism and the individual inhabitants of the pastoral lowlands in a way, which does not open any window of chance for a positive change. The pastoralists cannot but react on this threat by leaving the area with the cattle herd for seasons in succession with almost no way of return. In consequence of the given risky environment the people abandon their culturally inherited customs like the "culture of sharing" and try to survive as individual households and families, which is to be taken as a, if not fatal then at least highly damaging and un-just momentum of societal transformation.

The offer or challenge of decentralization and handing societal concerns, that is, governance, over to the people on the ground in the shape of committees does not offer any true chance of sustainable transformation. As the "state" governed committees do not open any room for recognitional and likewise procedural justice, and do neither acknowledge preexisting governance arrangements as well as worldviews, knowledges, or cultures nor safeguard feasible and acceptable ways of participation and decision making, these two dimensions of justice in societal transformation are not realized in any operational way. The process is to be taken as unsustainable.

Human rights focused critics of the Omo-Gibe basin, namely the Oakland Institute (2013), describe the effects of the upstream hydropower reservoir Gibe III and the land clearing for the Kuraz Sugar Development Project, that is, the forces at the regional scale, as a tragedy that has caused hunger and conflict. Again, and with relation to these forces, no justice in terms of recognition and installment procedures was and is granted to the people, the pastoralists on the ground, and as well constellations at this scale are to be taken as unsustainable.

Finally, the local scale in vision, the chapter turns against a too euphoric interpretation of the perspective of "transformations" in terms of climate services. Cattle feed from the residues of the sugar cane might offer some chance, maybe the only chance, for some partial loss compensation. But surely the more substantive aspects of a sustainable, which means a just handling of given situational settings, would need attention in the face of the claim for environmental and social justice:

- Which distributional effects will societal transformations have within the pastoralist societies? Who of the members of the pastoralist communities will be in the lead and govern the processes of feed storage, distribution and sale, and how will these subsequent leading social positions be legitimized?
- How will questions of societal justice, participation, and the distributive effects of climate change be addressed in social debates?
- Which changes in modes of governance and participation could be essential for any promising transformative processes?

References

Abbink J et al (2014) Lands of the future: transforming pastoral lands and livelihoods in eastern Africa. Working paper N0. 154. Max Planck Institute, Halle

Adger N, Paavola J, Huq S, Mace M (2006) Fairness in adaptation to climate change. MIT Press, Cambridge

Adger N et al (2009) Are there social limits to adaptation to climate change? Clim Chang 93:335–354

Appadurai A (1990) Disjuncture and difference in the global cultural economy. Theory Cult Soc 7(2–3):295–310

Ayele Z (2011) Local government in Ethiopia: still an apparatus of control? Law Democr Dev 15(1):1–27

Buffavand L (2016) 'The land does not like them': contesting dispossession in cosmological terms in Mela, South-West Ethiopia. J East Afr Stud 10(3):476–493

ESC (Ethiopian Sugar Corporation) (2014) Vision, Mission, Values. http://www.etsugar.gov.et/php/en/about/vision-mission-values

European Commission (2014) www.climate-services.org/about-us/what-are-climate-services

Famine Early Warning Systems Network: https://fews.net/east-africa/ethiopia

FDRE (2012) South Omo Villagization plan. Federal Democratic Republic of Ethiopia, Addis Ababa

Foucault M (1977) Dispositive der Macht. Über Sexualität, Wissen und Wahrheit. Merve-Verlag, Berlin

Hobold J et al (2019) Social-ecological change in the Omo-Turkana basin: a synthesis of current developments. Springer, without pages

Kamski B (2016) The Kuraz Sugar Project (KSDP) in Ethiopia: between 'sweet visions' and mounting challenges. J East Afr Stud 10(3):568–580

Kamski B (2019) Ethiopia Insight: Omo investors won't scrub away Kuraz's sugary stain. https://www.ethiopia-insight.com/2019/08/01
Lefort R (2012) Free market economy, 'developmental state' and party state hegemony in Ethiopia: the case of the 'model farmers'. Afr Stud 50(4):681–706
Mouffe C (2013) Agonistics: thinking the world politically. Verso, London/New York
Mueller-Mahn D, Everts J, Stephan Ch (2018) Riskscapes revisited – exploring the relationship between risk, space and practice. Erdkunde 72:197–213
Omo Turkana Research Network OTuRN (2016) Briefing note I – The Kuraz Sugar Development Project, Michigan State University, 1–6
Rancière J (1998) Disagreement: politics and philosophy. University of Minnesota Press, Minneapolis
Regessa A, Hizekiel Y, Korf B (2019) 'Civilizing' the pastoral frontier: land grabbing, dispossession and coercive agrarian development in Ethiopia, University of Zurich
Schatzki T (2002) The site of the social: a philosophical account of the constitution of social life and change. The Pennsylvania State University Press, University Park
Schatzki T (2012) A primer on practices. In: Higgs J, Barnett R, Billet S, Hutchings M, Trede F (eds) Practice-based education: perspectives and strategies. Sense Publishers, Rotterdam, pp 13–26
Stevenson EG, Buffavand L (2018) 'Do our bodies know their ways?' – Villagization, food-insecurity, and ill-being in Ethiopia's Lower Omo Valley. Afr Stud Rev 61:109–133
Swyngedouw E (2011) Interrogating post-democratization: reclaiming egalitarian political spaces. Polit Geogr 30(7):370–380
Temper L, Walter M, Rodriguez I, Kothari A, Turhan E (2018) A perspective on radical transformations to sustainability: resistances, movements and alternatives. Sustain Sci 13:747–764
The Oakland Institute (2013) OMO: local tribes under threat, a field report from the Omo Valley, Ethiopia. Oakland
Troeger S (2016a) Societal transformation, buzzy perspectives towards successful climate change adaptation: an appeal to caution, Chapter 19. In: Filho WL et al (eds) Handbook of climate change adaptation. Climate change management – implementing climate change adaptation in cities and communities, integrating strategies and educational approaches. Climate change management series. Springer, Cham, pp 353–365
Troeger S (2016b & 2018) 'Everything that is happening now is beyond our capacity' – Nyangatom livelihoods under threat, Chapter 19. In: Filho WL et al (eds) Climate change management series. Springer, pp 353–365. Republished in Cambridge University Press & UNESCO: Indigenous knowledge for climate change assessment and adaptation, Chapter 15. Cambridge, pp 214–226
Troeger S (2018) 'Committees', a promising institution in climate change communication and adaptation? Chapter 22. In: Leal W et al (eds) Handbook of climate change communication. Climate change management, vol 2. Springer, Cham, pp 341–355
Troeger S et al (2012) Failing seasons, ailing societies – climate change and adaptation in Ethiopia. Addis Ababa
Turton D (1985) Mursi response to drought: some lessons for relief and rehabilitation. Afr Aff 84 (336):331–346
Turton D (2010) The downstream impact. Presentation at the School of Oriental and African Studies, University of London, organized by the Royal Africa Society. www.mursi.org
www.climate-services.org/about-us/what-are-climate-services

7

Adaptation of Small-Scale Tea and Coffee Farmers in Kenya to Climate Change

Alice Nyawira Karuri

Contents

Abstract

The adverse effect of climate change on agriculture is well-documented and is a cause of concern for governments globally. In addition to concerns over food crop production, the economies of numerous developing countries rely heavily on cash crops. The coffee and tea sectors are key in Kenya's economy, contributing significantly to the gross domestic product, foreign exchange, and the direct or

A. N. Karuri (✉)
School of Humanities and Social Sciences, Strathmore University, Nairobi, Kenya
e-mail: akaruri@strathmore.edu

indirect employment of millions. Farmers engaged in the production of coffee and tea are predominantly small-scale farmers, with the majority farming on less than five acres. Climate change poses a threat to the production of these two crops and by extension to the economy of Kenya and the livelihood of farmers and those employed in these sectors. This study identifies the challenges posed by climate change in the tea and coffee sectors, the adaptation and mitigation measures identified, and the scope of their implementation. The production, processing, and marketing of tea and coffee in Kenya differs widely in terms of the institutions and institutional arrangements in the two sectors. This study will therefore analyze the role played by institutions in both sectors and how this affects climate change adaptation and mitigation measures by small-scale farmers.

Keywords

Kenya · Tea · Coffee · Small-scale farmers · Climate change adaptation · Institutions

Introduction

Agriculture in Kenya

The agriculture sector is key to Kenya's economy. It accounts for 65% of the export earnings and provides the livelihood of more than 80% of the population. The sector employs more than 40% of the total population and about 70% of the rural population. In 2018, it contributed 34.2% of Kenya's gross domestic product (GDP) and an additional 27% through linkages to other sectors such as manufacturing, distribution, and services (Food and Agriculture Organization (FAO) 2020; Kenya National Bureau of Statistics (KNBS) 2019; Ministry of Agriculture, Livestock and Fisheries (MALF) 2020). The climate of Kenya varies from tropical along the coast to arid in the interior. The weather in Kenya is generally sunny year-round, with the main rainy seasons being from March to May and from November to December (International Coffee Organization (ICO) 2019a). The topography rises from the coastal plains to the eastern edge of the East African Plateau and the Great Rift Valley. The highest altitude is in the central region and temperatures of 15 °C compared to the coastal region with temperatures of 29 °C (UNDP 2020). The Agriculture in Kenya is 98% rain fed and highly sensitive to changes in temperature and rainfall. Studies indicate there will be a 20% decrease in rainfall by the year 2030 (Government of Kenya (GoK) 2015). Temperatures are projected to increase 1.2–2.2 °C by 2050 in addition to increase in frequency and intensity of heavy rainfall, increase in severity of dry spells and duration of heat waves, and 16–42 cm rise in sea level (United States Agency for International Development (USAID) 2018). Since the early 1960s, both minimum and maximum temperatures have been increasing. The minimum temperature has risen generally by 0.7–2.0 °C and the maximum by 0.2–1.3 °C. There has been increased variability of rainfall from year to year and during the year. Extreme weather occurrences such as droughts and floods have become frequent and intense, leading to crop failures.

The adverse impacts of climate change are compounded by human factors such as illegal encroachments and settlements, logging and livestock grazing, which exacerbate deforestation, land degradation, and desertification. Forest cover in Kenya, for instance, has fallen from 12% in the 1960s to less than 2% in 2010. Kenya has a landmass of about 582,350 km^2 with 17% being arable while 83% consists of arid and semiarid land (ASAL) (GoK 2010). The combination of deforestation to open up croplands, the extension of agriculture onto land with low potential, and the use of more basic farming techniques and technologies due to cost and capacity barriers make the current agricultural system unsustainable in the long term (Republic of Kenya 2018). This necessitates that farmers engage in activities to adapt to and mitigate climate change. Adaptation refers to actions that minimize negative impacts of climate change, including the social, environmental, and economic impacts while mitigation refers to activities that reduce, prevent, or remove greenhouse gas (GHG) emissions and therefore reduce climate change. These measures require that farmers engage in sustainable agriculture, which is defined as farming in a responsible manner while enhancing profitability, well-being of the people, and the environment for now and the future (Cameron 2017).

The MALF oversees agriculture in the country. The Agriculture and Food Authority (AFA) is a government agency under MALF and is responsible for the development, regulation, and promotion of scheduled crops (ICO 2019b). AFA is comprised of several directorates that are specific to particular crops and include the Tea Directorate and the Coffee Directorate. Tea, horticulture, and coffee are Kenya's main agricultural exports. Tea and coffee are, however, unique as they are predominantly grown by small-scale farmers. This necessitates the formation of farmer organizations to benefit from economies of scale and to navigate the labor and capital-intensive process from production at the farm to the sale of the finished products.

Climate Change Policies and Regulations in Kenya

Kenya recognizes the importance of climate change action and has policies and plans to enact adaptation and mitigation measures. These include the National Climate Change Response Strategy (NCCRS) of 2010, National Policy on Climate Finance (2015), the Climate Change Act of 2016, the National Climate Change Action Plan (NCCAP) 2018–2022, and the National Adaptation Plan 2015–2030. The Environmental Management and Coordination Act (EMCA) of 1999 is the framework law on environmental management and conservation. Under this Act, EMCA established various institutions including the National Environmental Management Authority (NEMA). NEMA is the principal instrument of government charged with the implementation of all policies relating to the environment, and to exercise general supervision and coordination over all matters relating to the environment (National Environmental Management Authority 2020).

The NCCRS is the main document that guides the Kenya government's climate change agenda. The main focus of the strategy is to ensure that adaptation and mitigation measures are integrated in all government planning, budgeting, and development objectives. It prioritizes agriculture as one of the vulnerable sectors of the economy. The NCCRS established that the institutions in place to govern climate change affairs were inadequate and recommended that a comprehensive climate change policy and related legislation be put in place. The Climate Change Act was subsequently passed in 2016 and provides for a regulatory framework for an enhanced response to climate change. The Climate Change Act pertains to all sectors of the economy and to the national and county level government in the 47 counties of Kenya. It aims to mainstream climate change responses into development planning, provide incentives and obligations for private sector contributions in achieving low carbon climate resilient development, promote low carbon technologies, facilitate climate change research, and enhance cooperative climate change governance between the national government and county governments.

The Act obligates the Cabinet Secretary responsible for climate change affairs to formulate a five-year NCCAP. In accordance with the Act, the NCCAP represents the national mechanism through which climate change will be addressed in Kenya, including the implementation of the Nationally Determined Contributions (NDCs). Kenya submitted its NDCs in 2016. The NCCAP 2018–2022 provides mechanisms for mainstreaming climate change into all sectors of the economy and in the County Integrated Development Plans (CIDPs). The initiatives undertaken through the NCCAP include the scale up of renewable energy technologies, clean energy solutions, improved water resource management, sustainable forest management and tree planting, climate smart agriculture, and agroforestry (GoK 2015). It prescribes measures and mechanisms for climate change adaptation and mitigation, and the review and recommendation of duties of public and private bodies on climate change. Climate change duties refer to the statutory obligations conferred on public and private entities to implement climate change actions consistent with the national goal of low carbon climate resilient development. NEMA monitors, investigates, and reports on compliance with the assigned climate change duties. The Act also provided for financial provisions through the Climate Change Fund, which is a financing mechanism for priority climate change actions and interventions (Republic of Kenya 2016). Kenya also has a Climate Smart Agriculture Strategy for 2017–2026, which has been used as a source of input for the development of NCCAP 2018–2022 (Republic of Kenya 2018).

Tea Sector in Kenya

Kenya is the third largest producer of black tea globally after China and India, and is the world's largest exporter of black tea, contributing over 20% of total world exports. The sector is therefore significant to the global and national economy. By 2012 the sector accounted for 17% of total export earnings and 4 % of the national GDP (FAO 2015). In 2018, tea earnings amounted to Ksh.127.7 billion (KNBS 2019). Tea is grown on 236,000 ha with smallholders cultivating 142,000 ha and estates 93,000 ha. Tea production in 2018 was 493,000MT with 272,500MT from smallholders and 220,500 from estates (KNBS 2019). The tea sector contributes to environmental conservation through improved water infiltration, reduced surface erosion rates, and enhanced carbon sequestration.

Tea farming in Kenya is regulated by the Tea Act, Revised Edition 2012 [1960]. The Tea Directorate undertakes regulation and compliance of the tea industry, marketing and promotion of tea, the provision of technical and advisory services, and tea export guidelines. It facilitates research on all tea-related matters through the Tea Research Institute (TRI). All tea farmers are required to register with a factory as shareholders and to which they must supply the entirety of their production. Nearly all tea factories with small-scale farmer membership are shareholders in the Kenya Tea Development Agency (KTDA), a management company with several subsidiaries. Acreage per farmer varies from 0.25 to over 50 acres and most KTDA farmers grow on approximately half an acre of tea on average. Majority rely heavily on tea production for their livelihoods, which comprises over 60% of their total income. KTDA works directly with 611,000 farmers and indirectly impacts over four million people. A 2.5% management fee based on net price for services rendered is paid to KTDA Management Services and KTDA subsidiaries charge for their services separately. The government supports KTDA through guarantees for loans and support to extension services through the Ministry of Agriculture. The Kenya Tea Growers Association (KTGA) was established to promote the common interests of large-scale tea growers and is open to growers who maintain over 10 ha of tea. The large-scale tea sector, also referred to as tea plantation, includes both individual farmers and corporations, and accounts for about 40% of total tea production in Kenya (Kenya Tea Growers Association (KTGA) 2016). Challenges in the tea sector include declining prices, low yields, high production costs, low production diversification, low value addition, and a multiplicity of taxes and levies (Ngumo 2015).

The agro-climatic requirements of tea are temperatures ranging from 10 °C to 30 °C, ideally 0.5–10 degree slopes, elevations up to 2,000 m, acidic volcanic soils, well-distributed rainfall between the range of 1,200 and 1,400 mm annually, sufficient sunshine hours, and a mild climate. Tea in Kenya is grown in altitudes between 1,500 m and 2,700 m above sea level, receiving 1,200–1,400 mm of rainfall annually, which is spread throughout the year (FAO 2015). The agro-zones for tea production include the areas of Mount Kenya, Aberdare Range, Nyambene Hills, Mau Escarpment, Kericho Highlands, the hills of Nandi and Kisii, Mount Elgon, and Cherangani Hills (FAO 2015). Tea production in Kenya occurs all year round but the highest yields are in the rainy seasons in March–June and October–December. The tea supply is consistent throughout the year in both quantity and quality. Over 90% of tea from Kenya is handpicked, with only the top two leaves and a bud being picked for processing to ensure high quality. About 50 varieties of tea have been developed to suit the seven tea growing regions. The tea is grown without the use of agrochemicals as it is pest and disease free, and requires only fertilizer to replenish soil (TRFK 2010). Tea husbandry includes weeding, pruning, and fertilizer application. Pruning is ideally undertaken at the end of the peak-growing period, July to August, when the soil moisture is still adequate (East Africa Tea Trade Association 2020). The weather during this period is usually cold with light rains and enables pruning to occur without sun scorch.

Climate Change Challenges in the Tea Sector

Tea production is highly sensitive to changes in growing conditions. These conditions are expected to be impacted by climate change (Change 2015). Climate changes include inadequate rainfall, a larger soil water deficit, unpredictable rain patterns, and temperature rise (Ethical Trade Partnership (ETP) 2011). Climate change causes low productivity and poor quality in tea. Drought reduces the yields of tea while changes in the reliability and predictability of rainfall distribution and patterns have negative effects on tea yields and quality. Hailstorms and frost damage tea leaves and extreme temperatures suppress yields. High temperatures for example lead to decreased yields, reduced quality, high evaporation, reduced water content in the tea, dry weather pests, aggressive weed growth, weeds found in low country appearing in mid and up country poor bud break, bud scorch, stem and collar canker, wood root, leaf and bark desiccation. Excessive rain causes spread of fungal diseases, wet weather pests, and poor drainage in low-lying areas, heavy soil erosion that results in reduced water holding capacity, poor soil nutrients, and poor bud break and shoot development. Increase in extreme weather causes crop damage and failure due to events such as droughts, hail, storms, floods, frost, and landslides. Climate change also reduces productivity of subsistence crops which reduces food security (International Trade Center (ITC) 2014; Prematilake 2014). Due to climate change, the current areas of production are becoming unsuitable for tea production due to the risk of increasing temperature and increasing pests and diseases (ITC 2019; USAID 2018). A study by the International Center for Tropical Agriculture (CIAT) of climate change impacts on tea production in Kenya up to 2050, estimated that with increasing temperatures and rainfall, optimal areas for tea production will decrease, and production will have to shift to higher altitude areas, moving from around 1,500 m to 2,000 m above sea level (Ethical Tea Partnership 2011). A study by TRI indicates that in 2012, almost one-third of the harvest was lost (Omondi 2015). Incidences of severe and damaging frost that are attributed to climate change are becoming more common in Kenya. For example, the 2012 frost resulted in 30% tea crop loss in Nandi County (GoK 2015).

Adaptation and Mitigation Measures by Small-Scale Farmers in the Tea Sector

Tea planting is done through the planting of seedlings or more rarely, through the transplanting of a tree plant. Maturation of the tea seedling takes approximately 3 years, after which the tree plant continues production for decades. Adaptation to climate change can therefore not be carried out through adjusting the planting date. Adaptation measures can however be taken through adjusting the timing of tea husbandry activities such as pruning and fertilizer application, to climate change. Toward addressing issues on climate change, TRI is developing new technologies including environmental conservation efforts and development of improved tea varieties. TRI conducts research aimed at improving planting material, husbandry, yields, quality, and disease and pests control. It also provides advisory services to the growers on specific problems encountered in tea cultivation. The TRI has developed over 914 improved clones, of which 51 clones have been selected for consistent superiority in yield and quality. Thirteen of these clones yield between 5,000 and 8,000 kg of processed tea per hectare annually. These yield levels are some of the highest in the world and are three times the average yields of unimproved tea varieties. It has also developed a new tea clone – "Purple tea." According to the Kenya Agriculture and Livestock Research Organization (KALRO), the TRI began research on the "Purple Tea" cultivar over 25 years ago, and in 2011, the Tea Directorate began encouraging farmers to plant it. It is drought, frost, disease, and pest resistant. It has wide adaptability and is suitable for all designated tea-growing regions (KALRO 2020). Production of tea varieties such as purple tea, which is more resistant to climate variability than green tea, is a climate adaptation measure with the potential for higher income for tea farmers. Kimtai (2019) did a study aimed at establishing the rate of adoption of purple tea farming, determining the socioeconomic factors that hinders adoption and determining the role of purple tea farming for carbon sequestration. The purple variety had higher production than green tea and fetched higher prices. It was more resistant to drought, frost, hailstone, pests, and diseases. It was therefore highly rated for impacts of climate variability and change. Constraints to farming purple tea included availability of land, extension services were low particularly to farmers with the least acres of land (two acres), lack of training, poor access to credit, and limited market channels. The adoption of purple tea was however positively influenced by factors such as requiring little investment, higher income level, less risk on crop failure, and availability of labor.

Further adaption strategies include selection of the most suitable areas for tea growing, no expansion of new planting or replanting in low production areas, crop diversification in low production areas, efficient management of soil and water resources, catchment protection, riverbank protection, soil water conservation, crop insurance, use of drought tolerant cultivars, rainwater harvesting, and establishment of shade trees. Shade management particularly in the low and mid country reduces ambient temperature and prevents sun scorch. Other benefits of shade trees include carbon sequestration, improved net assimilation of tealeaf, reduced weed growth, additional organic matter from leaf litter and minimized wind damage, reduced frost, and reduced soil erosion (Ethical Tea Partnership 2011; Omondi 2015; Prematilake 2014). Tree planting also reduces frost and prevents soil erosion.

Several key partnerships enable the tea sector to adapt to and mitigate climate change. KTDA is involved in partnerships geared toward sustainable agriculture including climate change adaptation. One of its key partners has been Unilever, which is a multinational corporation with tea estates in Kenya and is also a major buyer of tea sold by KTDA. Unilever established a Sustainable Agriculture Programme in 1999. In 2007, it launched a partnership with the KTDA to enable Kenya's small-scale tea farmers acquire the certification standard set by the Sustainable Agriculture Network (SAN), which is a global coalition of environmental organizations. Factories and the KTDA trained farmers through Farmer Field Schools and Rainforest Alliance certification. Selected smallholder farmers are trained and they in turn instruct other farmers. Each lead farmer reaches an additional 300 farmers to ensure compliance with the SAN Standard required for Rainforest Alliance certification. KTDA in partnership with Unilever and IDH (the Dutch Sustainable Trade Initiative) has certified all the factories to the Sustainable Agriculture Network standards. Both farms and factories must meet the necessary requirements to receive Rainforest Alliance certification. By mid-2016, all of Kenya's smallholder tea farmers had met the Rainforest Alliance certification standards, and Unilever's Lipton brand was selling 100%-certified tea. Other major tea brands also began purchasing certified tea (Cameron 2017). The replicability, scale, and local leadership component of this system could serve as a model for developing participatory climate change adaptation plans (Moroge 2012). By March 2019, all 69 KTDA-managed tea factories were Rainforest Alliance certified and 21 tea factories were Fair Trade certified. Other certifications obtained by tea producers and factories included UTZ, Kosher, and Ethical Tea Partnership (ETP) (IISD (International Institute for Sustainable Development) 2019).

This success is attributed to several factors. Key among these is that Unilever, which has a large market share in Kenya, made the decision to purchase only tea that was Rainforest certified. Tea production is linked to global rather than domestic demand, with more than 95% of tea produced in Kenya is exported. Other factors include the strong regulation of the tea sector by the government including the registration and licensing of factories by the Tea Directorate. Kenya's tea industry is also highly concentrated both geographically and structurally with KTDA accounting for 60% of the market. KTDA's structure is highly integrated with strong links between farmers and factories (Cameron 2017). Unilever was also part of the development of the Cool Farm Tool, a calculator of GHG emissions that is freely available for use by farmers. In collaboration with the Kenyan government, it has been using it to quantify carbon within its plantations (Ellis et al. 2013).

The World Bank (WB) is also a key partner in climate adaptation in the tea sector. It announced during the March 2019 Nairobi Summit that an Emission Reduction Purchase Agreement (ERPA) would be signed between the Carbon Initiative for Development (Ci-Dev) trust fund, with WB acting as trustee, and KTDA Power Company Ltd. (KTDA Power). The contract aims to purchase carbon credits from small hydropower plants that provide power to 350,000 smallholder tea farmers and 39 of their regional tea factories in Kenya (Africa Times 2019). ETP, an alliance of tea packers working toward the sustainability of the tea sector, is another key partner. They started work in Kenya in 2010 and implemented a Climate Change Adaptation Program with GIZ, a German development agency. ETP utilized technology developed by its partners, such as Cafedirect Producers' Foundation, which developed a tool called WeFarm, an SMS platform for farmers (Budsock 2015). ETP also created a partnership with ITC and other nongovernmental organizations and founded a project to help farmers in climate change adaptation and mitigation. The program trains farmers and tea factory managers in carbon standards compliance, conservation and management of water, soil conservation, and use of biogas instead of wood. A partner factory, *Makomboki* Tea Factory that was using 2,000 cubic meters of wood per month as fuel for drying tea, changed to alternative energy sources. This included sawdust, rice husks, biomass, macadamia, and cashew nut shells and briquettes made from sawdust and rice husk. The new initiative saved more than 30,000 trees while lowering operational costs of that factory by 20% (Omondi 2015).

Multilateral partnerships are particularly effective in adaptation strategies. An example is the "Upscaling and Embedding Sustainability for Smallholder Tea Farmers," which is a collaborative initiative by KTDA Management Services (KTDA-MS), Unilever, and IDH – the Sustainable Trade Initiative. More than 85,000 (about 15% of 560,000 farmers) small-scale farmers had been trained by 2015 on Sustainable Agricultural Practices under a farmer field School (FFS) program (Cameron 2017). Another multilateral collaboration is between Germany, China, and the Food and Agriculture Organization (FAO), which was formed with the aim of promoting sustainable agricultural development and combating climate change in Kenya. This is through a project on carbon-neutral tea value chains. Germany through GIZ has been working with the KTDA, the Ethical Tea Partnership (ETP) among others in an integrated development partnership with the private sector to increase energy efficiency in all the factories to save GHG emissions and increase income of the farmers (Sino-German Center for Sustainable Development 2019). KTDA in 2015 signed a loan for Ksh.5.5 billion for the construction of seven small hydropower projects. The loan agreement was with the International Financial Corporation (IFC) which is a member of the World Bank, in partnership with other organizations including the Global Agriculture and Food Security Program (GAFSP) the French Development Institution (Proparco, the Netherlands Development Finance Company (FMO)). The expected reduction of reduce Kenya's carbon footprint from using hydropower is approximately 63,000 tons of carbon dioxide equivalent per year. On average, tea factories spend approximately Ksh. 30 million to Ksh. 65 million each per year on electricity (KTDA 2015). The hydropower plants have a total installed capacity of 16 megawatts and will provide captive power generation with the excess energy being sold to the state-owned power company. The farmers provided 35% equity on the loan via green leaf delivery, making it perhaps the first initiative in the world in which a farmer-owned institution is undertaking a renewable energy project of such a scale (IFC 2016).

KTDA Foundation, a nonprofit charity, also engages in environmental sustainability and climate change. Programs under climate change focus on promoting climate change mitigation, adaptation, and resilience building among smallholder tea farmers. In partnership with factories it has established over 31 indigenous, exotic, and fruit tree nurseries. Over 2.4 million trees have been planted in farms, major water catchment areas, and public forests. It is also undertaking an environmental conservation campaign for schools. Farmers are encouraged to adopt clean, renewable energy through promotion of and access to clean and renewable energy such as energy saving stoves, solar lighting products, and biomass. Adaptation measures include planting tea bushes along hill contours to reduce soil erosion. Seedlings from KTDA nurseries (commonly native species) are distributed throughout the local community for planting along farm boundaries, as buffers for waterways and forests, and to create small forest patches on farms. These native trees store carbon, stabilize the tea microclimate, and increase soil fertility. Planting native trees on steep slopes or degraded lands can also reduce vulnerability to heavy rains or prolonged droughts which creates resiliency to extreme weather. The KTDA is working to secure its own sources of sustainable fuel by acquiring land and encouraging its farmers to grow woodlots. Eucalyptus is commonly planted for fuelwood in the region. However, eucalyptus takes up large amounts of water, and KTDA and Rainforest Alliance are supporting farmers' efforts to replace eucalyptus in buffer zones with indigenous trees.

Although tea is on the "receiving end" of climate change, it also exacerbates it through deforestation. The common processing method of tea used in Kenya is the Cut Tear and Curl method, known as CTC. This involves the tea leaves going through a process of cutting, tearing, and curling, followed by oxidation and drying. The drying is often carried out using wood fuel due to the high price of electricity. Tea manufacturers are, however, gradually utilizing renewable energy alternatives such as electricity generation through micro-hydro plants to reduce energy costs and potentially generate further revenues by selling surplus electricity back to the national grid. Other mitigation measures include the use of biomass waste to power the water boiler systems. Gravity-powered ropeways are used by Finlay's, a multinational company, in some of its tea plantations. The KTDA also requires factories to acquire open land in order to plant seedlings and grow trees as a sustainable source of firewood. This saves money on the purchase of firewood or alternative fuels and could potentially generate revenues from carbon trading if the planted forests are managed sustainably. Varieties of tea other than Black CTC tea undergo a less emission-intensive process, as the wilting and CTC process are not required. Black CTC tea is manufactured by all 69 factories while only ten process black orthodox tea and four process other specialty tea.

Coffee Sector in Kenya

Kenyan coffee is grown on an estimated total area of 115,570 ha in 32 of 47 counties in the country. The sole type of coffee produced in the country is Arabica, which is planted during the rainy season from April to October with two harvest periods, April to June and October to December. Production is enabled by a combination of deep red volcanic soils, high altitude, rainfall, and moderate temperatures. The sole type of coffee grown in Kenya is Arabica. Coffee is grown in the high potential areas between 1,400 and 2,200 m above sea level, with temperature ranging from 15 °C to 24 °C, in red volcanic soils that are deep and well drained. Over 99% of Kenyan coffee is Arabica, whose main varieties are SL 28, SL 34, K7, Ruiru 11, Batian, and Blue Mountain (ICO 2019a). Coffee is an evergreen shrub and is therefore an important contributor to carbon sequestration, effective in stabilizing soils and permits the preservation of much of the original biodiversity in planted areas (ICO 2019b). The coffee sector contributes annually an average of US$230 million in foreign exchange earnings and is ranked as Kenya's fourth most important export, after horticulture, tourism, and tea. In 2018 coffee earnings amounted to Ksh. 14.8 billion. The value of coffee as a percentage of all export goods represented 5.5% in

2017, while its share of GDP was 0.22%. The coffee industry contributes an average of Ksh. 23 billion per year in foreign exchange earnings, ranking fourth after tourism, tea, and horticulture (ICO 2019a). Coffee is grown on 115,600 ha, 90,000 ha by smallholders and 25,000 ha by estates. Production was 41,400MT with 30,000MT from cooperatives and 11,000MT from estates (KNBS 2019). In 2017/18, the major importing countries of Kenyan coffee were Germany, the USA, Belgium, and the Republic of Korea, who imported about 58% of Kenya's coffee. Thirty-two of the 47 counties in Kenya are coffee producers (ICO 2019a).

Kenya has about 700,000 coffee farmers and about 99.63% have less than five acres. All coffee farmers with less than five acres are mandated to cooperative membership. There are between 500 coffee cooperatives (ICO 2019a) and 651 coffee cooperatives (KNBS 2019). Each coffee cooperative has factories where farmers deliver coffee berries for processing. At the factory, the coffee berries undergo pulping, which is a process of washing and drying, which results in "parchment." Parchment is then delivered to a coffee miller who further processes and grades the coffee, resulting in the "green" or "clean" coffee which is then sold by marketers at the Coffee Exchange through auctioning. Cooperatives also have the option of selling coffee directly to buyers without going through the auction. The Coffee Directorate is mandated to develop, promote, and regulate the coffee industry in Kenya. The Coffee Research Institute (CRI) conducts research in all areas of production, processing, and marketing of coffee. The global organization for coffee is the International Coffee Organization (ICO). It is comprised of member governments who represent 98% of world coffee production and 67% of world consumption. One of its objectives is to encourage members to develop a sustainable coffee sector in economic, social, and environmental terms (ICO 2019b).

Climate Change Challenges in the Coffee Sector

It is estimated that half the world's coffee-producing land will be unsuitable for coffee production by 2050 (Bunn et al. 2015; CIAT 2011). Other estimates indicate that the area unsuitable for production could be as high as 88% in Latin America (Worland 2018). Climate change caused by changing rainfall patterns and rising temperatures is affecting coffee production in several ways: directly through negative effects on the coffee plant and indirectly by altering the population dynamics and incidence of coffee pests and diseases (Jaramillo 2013). Rising temperatures will especially damage the Arabica bean, which accounts for about two-thirds of global coffee production, but whose production is limited to subtropical highlands in Brazil, Central America, and East Africa (Cameron 2017). This narrow region of the tropics is known as the coffee belt, and stretches from Central America to Sub-Saharan Africa to Asia (Worland, 2018). Rising temperatures will bring drought, increase the range of diseases, and kill large swaths of the insects that pollinate coffee plants (Worland 2018). Recent trends indicate that coffee growing is shifting from traditional optimal growing zones to higher altitudes. In traditional growing zones random flowering patterns and differences in berry growth stages has resulted in difficulties in disease and pest management, harvesting, and processing (Daily Nation 2012).

The effects of climate change in Kenya include unreliable and erratic rains with shorter seasons (ICO 2019b). Changing rainfall patterns affect the flowering of the coffee plants which impacts the whole production cycle. The altered flowering pattern with coffee berries at different stages of maturity poses a challenge in disease management, insect management, and harvesting (Ethical Tea Partnership 2011; Mugo 2016; Mwaura 2010). Extremely heavy rains lead to higher erosion levels, resulting in loss of soils, leaching of nutrients, and consequent soil infertility. On the other hand, the dynamics of incidence of coffee pests and their management are evolving rapidly due to changing climatic conditions. The changing environment is also posing challenges to patterns of cherry ripening and drying of parchment because of unpredictable rainfall patterns (ICO 2019b). Changing rainfall patterns cause uncertainty regarding the timing of fertilizer application and the drying of parchment. Most coffee growing zones in central Kenya, particularly Kiambu and Murang'a, are no longer suitable for the crop due to rising temperatures (Kamau 2017). Intermittent rainfall in the 2007/08 crop year caused a severe episode of Coffee Berry Disease that cut Kenyan output by 23% to 42,000 MT. This happened because farmers were not able to spray the crop on time (Mwaura 2010). Climate variability and its effects is however not a "new" problem, as evidenced by research shortly after Kenya's independence. This is illustrated by a 1969 journal article by Nutman and Roberts, titled "*Climatic conditions in relation to the spread of coffee berry disease since 1962 in the in the East Rift Districts of Kenya.*" (F.J. Nutman & F.M. Roberts 1969)

Adaptation and Mitigation Measures by Small-Scale Farmers in the Coffee Sector

The Coffee Directorate in collaboration with stakeholders provides capacity building to the counties' agricultural staff and other value chain players. The collaborating private agencies include Technoserve, Solidaridad, certification bodies such as UTZ, 4C, and Fairtrade, and management services providers. CRI develops technologies, releases new coffee varieties, and carries out research on disease and pest management, while the Ministry of Agriculture sets policy guidelines (ICO 2019b). The Coffee Research Foundation (currently CRI) started a program in 2012 to help farmers reverse the effects of global warming and boost coffee production. Farmer sensitization clinics were held and farmers were encouraged to plant indigenous trees to provide shade and to practice water harvesting (Daily Nation 2012). The CRI has made various recommendations on approaches toward environmentally sustainable coffee production systems. One approach is integrated farming, in which mulching, conservation agriculture, organic fertilizers, and use of bio-stimulants are recommended. Other measures include the use of suitable shade trees and the adjustment of spraying programs to cope with changing trends in the manifestation of coffee pests and diseases (ICO 2019a). Most Kenyan coffee is grown without shade, but shaded coffee is becoming increasingly popular in order to mitigate the effects of climate change, although quantifying the area under shade has not been done. Research is ongoing to determine the appropriate shade trees (ICO 2019a). Research shows that biodiverse shaded coffee is far more resilient and productive than coffee grown in monoculture. Shading coffee therefore improves the resilience of agro-ecosystems. Shade trees protect plants from microclimate variability, the effects of lower precipitation and reduced soil water availability, and reduce high solar radiation. It also improves soil fertility, protects coffee from insect pests, and provides economic benefits for farmers (Jaramillo 2013).

The need to develop disease-resistant coffee varieties was felt in early 1971 and breeding programs were initiated whose optimal outcome was the cultivar Ruiru 11 (Njoroge 1991). The CRI currently has developed two improved varieties of coffee – Ruiru11 and Batian. The improved varieties are resistant to Coffee Berry Disease and Leaf Rust Disease, thus lowering the use of agrochemicals and reducing production costs. The CRI estimated that the production cost of the traditional variety is four to five times more than Ruiru 11 and Batian. Over 300,000 farmers are estimated to have planted the new varieties (ICO 2019a). The traditional variety of coffee grown is SL-28. Other Arabica varieties include SL-34, K7, and Blue Mountain. A study of the coffee sector in Nyeri County showed that 50% of farmers in certified coffee cooperatives and 57% in the noncertified cooperatives had only the traditional SL variety of coffee on their farm. An additional 45% in the certified and 39 per in the noncertified had other varieties in addition to the SL (Okech 2019). It therefore means that less than 5% had the disease and drought resistant varieties exclusively on their farms. Incentives therefore need to be provided to encourage planting of new varieties, beyond the step of providing free seedlings.

The International Coffee Organization (ICO) has pilot projects in Africa and Latin America to address climate change by assisting coffee farmers use environmentally friendly technologies. These include building the capacity of institutions, improving access to credit and risk management mechanisms, reducing vulnerability to income volatility, and promoting gender equality. It also engages in long- and short-term adaptation strategies as well as mitigation strategies (ICO 2019b). Several organizations including private companies involved in coffee production and marketing have taken the lead in climate change adaptation and mitigation in the coffee sector. The World Coffee Research (WCR), a consortium supported by major coffee retailers, distributors, and exporters, has an $18 million coffee-monitoring program that covers 1,100 farms in 20 countries including Kenya. It conducts farmer training, provides technical assistance, and is testing coffee varieties and adaptive farming methods (Worland 2018). Since 2013 Starbucks has support centers in nine countries and a 10-year, $500 million investment fund that supports sustainability programs, including adaptation training for farmers and the testing of new coffee varieties (Worland 2018). Sangana Commodities Ltd and GIZ implemented a three-year project creating a link between coffee smallholders and carbon markets, and developing a verifiable and voluntary climate change module, which can be integrated into the existing 4C's standard (ETP 2011).

Sustainability standards in the coffee sector are geared toward sustainable agricultural practices, which include adaptation and mitigation measures. Certification in the coffee sector is, however, not as extensive as in the tea sector. In Nyeri County, which is a leading coffee producer, Fairtrade was the most common certification with at least 12 of the 23 cooperatives having received this certification. The earliest reported Fairtrade certification in the cooperatives was from 2006. At least seven cooperatives had obtained Rainforest Alliance certification and four had acquired 4C certification. Benefits of certification included payment of premiums and improved quality and quantity of coffee. Challenges included difficulty of maintaining certification as it was involving and expensive. This is illustrated by the example of an estimated Ksh. 400,000 for certification and renewal, with additional costs of up to Ksh. 1,000,000. Enforcing requirements as well as nonconformity to requirements was also costly. Certification organizations also did not source for markets and the cooperatives therefore used the conventional marketing channels, which were viewed as opaque (Okech 2019).

Recommendations and Conclusion

Tea and coffee are essential to farmers and other Kenyans as a source of livelihood, to the government as a contributor to GDP and foreign exchange and to the commodity chain players such as marketers and retailers. Tea and coffee also contribute significantly to climate change mitigation. Climate change, however, threatens to disrupt the production of these crops and by extension the economy of Kenya and the livelihoods of those who depend on it. Continued and increased uptake of climate adaptation and mitigation measures is critical for the sustainable farming of tea and coffee. Climate change can worsen the socioeconomic condition of farmers and conversely fragile socioeconomic conditions can exacerbate climate change. Various measures can be taken to increase the adaptive capacity of small-scale tea and coffee farmers.

Farmer Empowerment

Farmers are rational economic actors and farming has to be profitable for adaptation and mitigation measures to be implemented. Climate adaptation and mitigation activities such as uptake of improved crop varieties are carried out in view of perceived economic benefits and particularly for the most vulnerable, in view of immediate economic benefits. Activities such as the adaptation of improved crop varieties can be neglected due to small land size and the resultant "lost" while the new variety matures to the point of harvest. Other measures such as water harvesting may require capital expenditures whereby the capital is either unavailable or the returns to investment are deemed uneconomical. Interventions that deal with commodity chain weaknesses that reduce profitability are critical. Profitability of farming is the largest incentive to adopting sustainable agricultural practices. Increasing the adaptive capacity of tea farmers therefore requires farmers' economic empowerment. It is essential that stakeholders provide incentives for the implementation of measures whose returns are long-term or not tangible. This could include measures such as insurance for climate-related crop losses and a minimum guaranteed price per kilogram. Farmers also need to be politically empowered as decisions in farmer organizations are often made by elected representatives. Civic awareness is therefore an integral component of farmer empowerment. Social empowerment majorly revolves around gender. Men predominantly own land but women provide 60% of the labor on the farms and in the wet mills (ICO 2019b). Women therefore need to be major actors in activities involving climate change adaptation and mitigation. When they are not empowered, especially in terms of access to resources, various adaption and mitigation activities may not be implemented.

Strengthening of Institutions

The common features of the tea and coffee sector in Kenya is that both are cash crops, grown majorly for export and production is predominantly by small-scale farmers. Profitable production therefore requires that farmers organize so as to share costs. In the tea sector, the KTDA is the primary agency through which farmers produce process and market their tea. In the coffee sector, coffee cooperatives are the primary and mandated vehicle for production, marketing, and processing. However, there are coffee farmers with more than five acres who choose to remain in cooperatives due to economies of scale. Institutional capacity therefore has a large influence on farmers' activities, including those related to climate change adaption and mitigation. Lack of institutional capacity in farmer-owned organizations or institutions that support farmers can reduce productivity and profits and therefore constrain farmers' adaptive capacity. The average age of tea and coffee farmers is over 50 years and this further constrains the availability of alternative livelihoods. Unlike the tea sector where almost all small-scale tea farmers operate within the institutional structure of KTDA, coffee cooperatives are not homogeneous. For example, coffee payments for the 2017/18 year varied from Ksh. 9 to Ksh. 105/kg of coffee cherry (exchange rate in this period was approximately one USD to Ksh. 100–103). While there are climatic and soil-type differences, the differing payments were also within the same region. In Nyeri County, which is a leading coffee producer, the payment range was between Ksh. 12.75 and Ksh. 105. This points to institutional constraints at the cooperative level, although constraints occur across the coffee commodity chain. The challenges facing the coffee sector are well documented. Toward this end, a Coffee Taskforce was created to investigate the challenges and make recommendations for sectoral improvement. The tea sector has also encountered various institutional challenges. In January 2020, the president directed that the KTDA be restructured for the benefit of tea farmers (Cheruiyot 2020). The organizations that farmers engage in and the institutions within which they operate can either facilitate or constrain the activities of farmers, including those pertaining to climate change adaptation and mitigation.

Collaborations and Partnerships

Climate change adaptation and mitigation requires the long-term involvement of all actors in the commodity chain including the consumers. Stakeholder mapping is necessary, whereby a list of relevant groups, organizations, and people who can collaborate in climate change adaptation and mitigation can be invited to collaborate towards these measures. These stakeholders include the International Coffee Organization, the Tea and Coffee directorates, agricultural institutions and departments, producer organizations, certification organizations, technical support providers, financial institutions, and supply chain actors including processors, marketers, and retailers. Research, information, and best practices on sustainable agricultural practices and on measures to streamline the commodity chain can be enhanced and widely adopted through collaboration.

Certification

The 2017 SAN Standard aims to support farmers in advancing sustainable livelihoods, improving farm productivity, and becoming more resilient to climate change. Changes of note include climate-smart agricultural practices. These are built into the standard to help farmers address climate change risks. The effect of irregular rainfall, changing temperatures, and related increased pest and disease attacks can be reduced through soil conservation, water-use efficiency, and the conservation and restoration of natural ecosystems. The Standard is built on principles of sustainable farming including biodiversity conservation, improved livelihoods and human well-being, natural resource conservation, effective planning, and farm management systems (Rainforest Alliance 2019). The SAN Climate Module is an add-on for voluntary verification within the existing Sustainable Agriculture Network certification system. Farmers who achieve compliance with the module will be able to assess the risks posed by climate change to their farms and communities, analyze their practices to quantify and reduce GHG emissions, and increase the carbon levels stored on their farms through the restoration of degraded lands, reforestation, and improved soil conservation while also being able to better adapt to altered growing seasons and other conditions (Sustainable Agriculture Network 2011).

Fairtrade aims to help farmers become more resilient to climate change while giving consumers, retailers, and traders the opportunity to reduce their carbon footprint. Farmers can spend the Fairtrade Premium on climate change adaptation projects such as tree planting, irrigation, crop diversification, and clean energy. Farming communities can also benefit from access to carbon finance, which can be used in mitigation or adaptation activities (Fairtrade International 2015). While certification has numerous benefits, acquiring and maintaining of certification by coffee cooperatives and estates requires the collaboration and financial support of stakeholders.

Rainforest Alliance and Fairtrade are the major certification bodies in the coffee and tea sector in Kenya. Majority of the small-scale tea farmers in Kenya comply with the SAN Standard and have received Rainforest Alliance. This extensive certification can be largely attributed to the institutional arrangement of the tea sector. KTDA acts as the management agent for 69 factories, which comprise the vast majority of small-scale tea farmers in Kenya. In the coffee sector, however, more than 700,000 farmers are members of an estimated 500–650 coffee cooperatives. Certification depends on various factors including membership numbers, productivity levels, number of factories, cooperative leadership, and relationships with a multiplicity of commodity chain actors. While some cooperatives are certified by multiple certification bodies, others do not comply with any sustainability standard. Certification ensures the implementation of climate change adaptation and mitigation measures. Acquiring and complying with sustainability standards is therefore key for small-scale farmers.

References

Africa Times (2019) World Bank commits $22.5 billion more to Africa for climate change. Africa Times. Retrieved 10 Dec 2019 from https://africatimes.com/2019/03/15/world-bank-commits-22-5-billion-more-to-africa-for-climate-change/

Budsock A (2015) Ethical tea partnership – adapting smallholders to climate change. Retrieved 10 Dec 2019 from https://impakter.com/ethical-tea/

Bunn, Laderach, Rivera and Kirschke (2015) A bitter cup: climate change profile of global production of Arabica and Robusta coffee Climate Change 129:89–101

Cameron B (2017) Brewing a sustainable future: certifying Kenya's smallholder tea farmers, 2007–2017. Woodrow Wilson School of Public and International Affairs, Innovations for successful societies. Princeton University. Retrieved 10 Dec 2019

Change K (2015) World tea production and trade current and future development, Food and agriculture organization, trade and markets division. FAO, Rome

Cheruiyot K (2020) Uhuru orders for KTDA restructuring to save tea farmers. The Star. Retrieved 22 Jan 2020 from https://www.the-star.co.ke/news/2020-01-14-uhuru-orders-for-ktda-restructuring-to-save-tea-farmers/

Daily Nation (2012) New plan to help farmers reverse effects of climate change. Daily Nation. Kenya. Retrieved 10 Dec 2019 from https://reliefweb.int/report/kenya/new-plan-help-farmers-reverse-effects-climate-change

East Africa Tea Trade Association (2020) Tea care and husbandry management. Retrieved from eatta.com

Ellis K, Lemma A, Mutimba S, Wanyoike R (2013) Low carbon competitiveness in Kenya. Policy brief, Overseas Development Institute. Retrieved 17 Jan 2020

Ethical Tea Partnership (2011) Climate change adaptation in the Kenyan tea sector. Report from the adaptation workshop held in Kericho on the 16th of May 2011. Retrieved from ethicalpartnership.org

Fairtrade International (2015) Climate change. Retrieved 2020 from fairtrade.net: https://www.fairtrade.net/issue/climate-change

FAO (Food and Agricultural Organization) (2020) Kenya at a glance. Retrieved from www.fao.org: http://www.fao.org/kenya/fao-in-kenya/kenya-at-a-glance/en/

FAO (Food and Agriculture Organization of the United Nations) (2015) Kenya's tea sector under climate change. An impact assessment and formulation of a climate-smart strategy, by Elbehri A, Cheserek B, Azapagic A, Raes D, Mwale M, Nyengena J, Kiprono P, Ambasa C. Rome, Italy

Government of Kenya (2015) Addressing climate change: success stories from Kenya. Ministry of Environment, Natural Resources and Regional Development Authorities. Government of Kenya. Retrieved 17 Jan 2020

Government of Kenya (2010) National climate change response strategy. Executive Brief. Retrieved from environment.go.ke: http://www.environment.go.ke/wp-content/documents/complete%20nccrs%20executive%20brief.pdf

IFC (2016) Kenya's tea farmers taste benefits of hydropower. International Finance Corporation. Retrieved 9 Mar 2020 from ifc.org: https://www.ifc.org/wps/wcm/connect/news_ext_content/ifc_external_corporate_site/news+and+events/news/impact-stories/tea-farmers-taste-the-benefits-of-hydropower

IISD (International Institute for Sustainable Development) (2019) Sustainability and voluntary certification in the Kenyan tea sector. Developing an action plan to address opportunities and challenges. Report of the Meeting Held in Nairobi, Kenya, March 6, 2019. Retrieved 29 Apr 2020 from iisd.org

International Coffee Organization (2019a) Country coffee profile: Kenya. International Coffee Organization. International Coffee Council. Retrieved 17 Jan 2020 from http://www.ico.org/documents/cy2018-19/icc-124-7e-profile-kenya.pdf

International Coffee Organization (2019b) Developing a sustainable coffee economy. Retrieved 17 Jan 2020 from http://www.ico.org/sustaindev_e.asp

International Trade Center (2014) Mitigating climate change in the tea sector. ITC, Geneva. Retrieved from: intracen.org on 17 Jan 2020

Jaramillo JSM-O (2013) Climate change or urbanization? Impacts on a traditional coffee production system in East Africa over the last 80 years. (Hart JP, ed). PLoS ONE 8(1). Retrieved 17 Jan 2020 from https://journals.plos.org/plosone/article?id=10.1371/journal.pone.0051815

KALRO (2020) Purple tea. Retrieved from kalro.org: http://www.kalro.org/tea/?q=Purple%20Tea

Kamau G (2017) Coffee takes beating from rising global temperature. The East African. Retrieved 17 Jan 2020 from https://theeastafrican.co.ke/business/climate-change-effect-on-coffee/2560-3998982-giionx/index.html

Kenya National Bureau of Statistics (KNBS) (2019) Economic survey 2019. Economic survey, Kenya National Bureau of Statistics. Retrieved 10 Dec 2019 from www.knbs.or.ke

Kenya Tea Growers Association (KTGA) (2016) Kenya tea growers association. Retrieved 23 Apr 2020 from Kenya Tea Growers Association: ktga.or.ke

Kimtai R (2019) Adoption of purple tea farming as a coping mechanism to climate variability and change. A case study of Kericho County, Kenya. Masters thesis, Jomo Kenyatta University of Agriculture and Technology, Environmental Legislation and Management. Retrieved 17 Jan 2020

KTDA (2015) KTDA Signs Ksh. 5.5 Billion loan agreement to construct seven small hydropower projects. Retrieved 9 Mar 2020 from ktdateas.com: https://www.ktdateas.com/index.php/blogs/item/49-ktda-signs-ksh-5-5-billion-loan-agreement-to-construct-seven-small-hydropower-projects.html

Ministry of Agriculture, Livestock and Fisheries (2020) Towards sustainable agricultural transformation and food security in Kenya 2019–2029. Government of Kenya. Retrieved 10 Mar 2020 from http://www.kilimo.go.ke/wp-content/uploads/2019/01/ASTGS-Full-Version.pdf

Moroge M (2012) In Kenya, tea growers adapt to climate change. Retrieved 10 Mar 2020 from rainforest-alliance.org: https://www.rainforest-alliance.org/pictures/kenya-tea-growers

Mugo I (2016) Global warming takes its toll on coffee production. Daily Nation. Kenya. Retrieved 10 Dec 2019 from https://www.nation.co.ke/news/Global-warming-takes-its-toll-on-coffee-production/1056-3461704-avbq99z/index.html

Mwaura HN (2010) Climate change affecting Kenya's coffee output. Retrieved 10 Dec 2019 from https://www.reuters.com/article/us-kenya-coffee-climate/climate-change-affecting-kenyas-coffee-output-idUSTRE61A0WA20100211

National Environmental Management Authority (2020) Environmental Act (EMCA). Retrieved from nema.go.ke: http://www.nema.go.ke/index.php?option=com_content&view=article&id=24&Itemid=163

Ngumo D (2015) My experience in the tea sector. Retrieved from karuspace.karu.ac.ke

Njoroge J (1991) Management of hybrid Ruiru 11 [eleven] Arabica coffee – a review. Coffee Research Foundation, Ruiru. Retrieved 10 Dec 2019 from www.kainet.ne.ke/documents/management-hybrid-ruiru-11-eleven-arabica-coffee-review-1

Nutman FJ, Roberts FM (1969) Climatic Conditions in Relation to the Spread of Coffee Berry Disease Since 1962 in the East Rift Districts of Kenya. East African Agricultural and Forestry Journal 35 (2):118–127

Okech AN (2019) Producer institutional arrangements in Kenya's coffee sector and their effect on economic benefits to farmers. PhD thesis. Department of Development Studies. Jomo Kenyatta University of Agriculture and Technology, Juja

Omondi D (2015) Climate change bites Kenyan tear farmers. Voices2Paris. Retrieved 10 Dec 2019

Prematilake K (2014) Climate change adaptation strategies for tea plantations. Tea Research Institute of Sri Lanka, Agronomy Division. Retrieved 17 Jan 2020

Rainforest Alliance (2019) What does rainforest alliance certified mean? Retrieved 11 Mar 2020 from rainforest-alliance.org: https://www.rainforest-alliance.org/faqs/what-does-rainforest-alliance-certified-mean

Republic of Kenya (2016) The climate change act, 2016. Kenya gazette supplement. Government Printer, Nairobi

Republic of Kenya (2018) National climate change action plan 2018–2022. Ministry of Environment and Forestry. Government of Kenya

Sino-German Center for Sustainable Development (2019) Carbon-neutral tea value chains – Germany, China and the FAO Join Forces with Kenya. Retrieved 9 Mar 2020 from https://sg-csd.org: https://sg-csd.org/news_events/20190328/

Sustainable Agriculture Network (2011) SAN climate module criteria for mitigation and adaptation to climate change. Retrieved from www.rainforest-alliance.org: https://www.rainforest-alliance.org/lang/sites/default/files/site-documents/climate/documents/SAN-Climate-Module-January2011.pdf

UNDP (2020) Climate change adaptation. Kenya. Retrieved from adaptation-undp.org

USAID (2018) Climate risk profile Kenya. Retrieved from: climatelinks.org on 17 Jan 2020

Worland J (2018, July 2) Your morning cup of coffee is in danger. Can the industry adapt in time? TIME. Retrieved 17 Jan 2020 from http://time.com/5318245/coffee-industry-climate-change/

8

Brachiaria Grass for Climate Resilient and Sustainable Livestock Production in Kenya

D. M. G. Njarui, M. Gatheru and S. R. Ghimire

Contents

D. M. G. Njarui (✉) · M. Gatheru
Kenya Agricultural and Livestock Research Organization (KALRO), Katumani, Kenya
e-mail: donaldnjarui@yahoo.com

S. R. Ghimire
The Biosciences eastern and central Africa – International Livestock Research Institute (BecA-ILRI) Hub, Nairobi, Kenya

Abstract

Brachiaria grass is a "climate smart" forage that produces high amount of palatable and nutritious biomass for livestock and performs well in infertile soils, sequesters carbon in soil, and provides several environmental benefits. The objective of the study was to validate the productivity of *Brachiaria* grass and upscale the suitable cultivars for improved livestock feed resources in Kenya. We assume integrating *Brachiaria* grass into mixed crop-livestock system will enhance feed availability and livestock productivity, leading to increased food and nutrition security. Farmer participatory approach was adopted to evaluate and promote four *Brachiaria* grass (*Brachiaria decumbens* cv. Basilisk, *B. brizantha* cvs. Xaraes, Piata, and MG-4) in the Central Highland and Eastern Midland of Kenya. The extension/advisory approaches used to promote *Brachiaria* grass cultivars included field days, village knowledge centres, agricultural shows, posters, and linkages with other institutions through multi-actor platform established under the InnovAfrica project. Generally, *Brachiaria* grass cultivars were more productive than the control (Rhodes grass) in most harvests reaching peak of 5.1–7.7 t/ha in the fifth harvest. For Rhodes grass, DM was less than 4 t/ha in all harvest and died by sixth harvest. Similarly, based on farmers' evaluation using phenotypic traits, the *Brachiaria* grass cultivars had higher score than Rhodes grass except cv. Piata. The mean score ranged from 2.75 to 3.19 for *Brachiaria* cultivars, while for Rhodes the mean score was 2.63. Within 2 years of intervention, over 4000 farmers in the 2 project sites and additional 1500 farmers from other parts of the country have planted the *Brachiaria* grass. The demand for *Brachiaria* grass seeds is increasing due to benefits gained, e.g., increased milk production from dairy cattle fed on the grass. Our study will quantify the associated benefits from cultivation of *Brachiaria* grass with respect to a set of ecological, food and nutrition security, and social-economic indicators.

Introduction

There is wide empirical evidence that livestock plays an important economic role in Kenya. The livestock subsector contributes about 12% to the national Gross Domestic Product in the country (RoK 2019). Livestock serves multiple purposes that include food and nutritional security. Livestock is sold to generate income; provide manure and draft power for crop production, skins, and transport; and serve in social-cultural function and source of prestige within the pastoral society. The effects of climate change have challenged the sustainability of livestock sector. The changing pattern and frequency of extreme climatic conditions such as droughts and floods have great impacts on livestock and the associated livelihoods (Freier et al. 2012; Schilling et al. 2012). Extreme climatic events such as droughts and heavy rains are expected to become more frequent in the Horn of Africa and part of East Africa (Christensen et al. 2007; World Bank 2013). Climate change leads to reductions in livestock productivity by indirectly compromising the availability of forages (Martin et al. 2016). The impacts on forage availability and quality may include changes in herbage growth, changes in composition of vegetation, and overall changes in herbage quality (Thornton et al. 2009).

In the mixed crop-livestock production systems where dairy farming is practiced, several forage innovations have been in use to alleviate livestock feed shortage. Napier grass is the most widely cultivated forage in the humid and subhumid regions of Kenya, and over 90% of dairy farmers depend on it, but it is threatened by smut and stunting diseases. These diseases have been reported to cause damage ranging from 5% to 90% (Lusweti et al. 2004). There is need to introduce climate smart technologies to improve resilience and sustainable livestock systems to climate change. One option is the *Brachiaria* grass which is a "climate smart" and increases livestock productivity (Muinga et al. 2016; Ngila et al. 2016).

Brachiaria is one of the important tropical forage grasses of African origin. It is widely cultivated in South America, Australia, and East Asia and has demonstrated success in transformation of beef and dairy industries (Ghimire et al. 2015). The perennial *Brachiaria* grass species produce high tonnage of foliage biomass, possess large root systems, sequester carbon into soils, is adapted to drought and low fertility soils, and provide several environmental benefits and ecosystem services (Djikeng et al. 2014; Njarui et al. 2016).

Although native to Eastern and Central Africa, its use as livestock feed has been extremely limited in the region including Kenya. Improved *Brachiaria* grass cultivars identified in evaluations conducted in Eastern Midlands (EM) region of Kenya can bridge the gap on livestock feed scarcity. However, there is limited information available on the productivity of different *Brachiaria* cultivars in different agroecological zones (AEZ) of Kenya. For instance, only a limited number of farmers are aware on benefits of *Brachiaria* grass to livestock productivity. The dry matter (DM) yield of *Brachiaria* grass varies in different AEZ and is influenced by a range of factors including cultivars, moisture, soil fertility, pest and disease, and management options. To achieve a sustainable impact of *Brachiaria* grass in Kenyan livestock sector, a wide-scale adoption of this grass is necessary. This requires selection of right cultivars for a given production environment prior to integration in the farming system, and engaging farmers in the cultivar selection would be crucial for adoption and scaling of the technology. The objective of the study was to validate the productivity of *Brachiaria* grass and upscale the suitable cultivars for improved livestock feed resources in Kenya.

Methodology

Description of the Study Area

The study was carried out in Kirinyaga County in Central Highlands (CH) and Kangundo subcounty in EM (Fig. 1). These sites differ in agroecological characteristics and diversity in farming system. *Brachiaria* grass had been tested and found to be suitable in the EM and hence was upscaled among the smallholder crop-livestock farmers who keep dairy cattle for commercial production. We adopted farmer participatory approach to evaluate and promote *Brachiaria* technologies in the CH and EM of Kenya using a novel extension approach – i.e., the Village Knowledge Centre (VKC) supported by innovative institutional approach – multi-actor platforms (MAP) that was established and being operational under the InnovAfrica project (www.innovafrica.eu). On the other hand, *Brachiaria* grasses were evaluated in the CH since no research had been conducted on this grass in the region. The biophysical, farming system and other features of study sites are described below.

Kirinyaga County: Kirinyaga is one of the counties in the CH of Kenya. It is located between latitudes 0° 1′ and 0° 40′ South and longitudes 37° and 38° East and

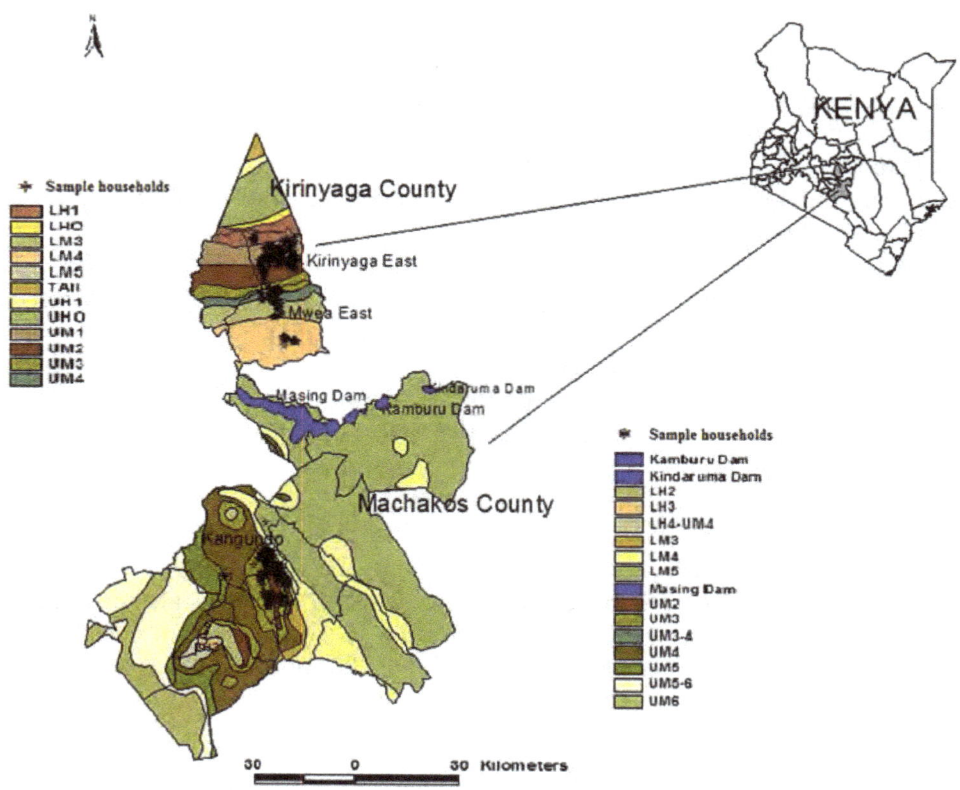

Fig. 1 Map showing project study sites

covers an area of 1478 km^2. The altitude ranges from 1200 to 1700 m above sea level (asl), but in the northern side, it rises to 5200 m at the top of Mt. Kenya. The major soil is Humic Nitosols (Jaetzold et al. 2006). According to 2009 census, the population stood at 528,054 persons comprising 154,220 households (KNBS 2010).

Rainfall is bimodal, the long rains occurring from March and May and the short rains from October to December with average annual rainfall from 1200 to 1500 mm, which is well distributed and reliable for forage production. Temperature ranges from 12 °C to 27 °C (Jaetzold et al. 2006). Farming is predominantly mixed crop-livestock system. The main cash crop are rice, tea, and coffee, while maize, beans, bananas, and sweet potatoes are the main food crops but are often sold to generate income. Major livestock are cattle, goats, sheep, and poultry. Dairy farming is an important enterprise and is based on exotic cattle and their crosses, mainly under stall feeding system. Napier grass is the main feed resource, while banana stems, sweet potatoes, and crop residues are fed during the dry season.

The *Brachiaria* grass validation trial was conducted at Kamweti Agricultural Training Centre (ATC) in Kirinyaga Country. Kamweti ATC lies on the southern slopes of Mt. Kenya at an altitude of 1755 m asl. The soil characteristics from samples taken (0–30 cm depth) indicated that the texture of the soil was sandy clay, with average pH of 4.87, total nitrogen content (0.23%) and organic carbon about 2.5%.

Kangundo subcounty: Kangundo, a subcounties of Machakos County, lies between latitudes 0° 45′ and 1° 31′ South and longitudes 36° 45′ and 37° 45′ East With altitude range of 800–1600 m asl. This subcounty has an area of 177 km^2 with population of about 94,367 (KNBS 2010). The predominant soils are Luvisols, Acrisols, and Ferralsols derived from the Precambrian "basement-complex" rocks (Simpson et al. 1996). They are often shallow, contain low organic matter and high sand content (Kusewa and Guiragossion 1989), but are well drained.

Rainfall pattern is similar to Kirinyaga County but is lower and ranges from 800 to 1050 mm annually. Inter-seasonal rainfall variation is large with coefficient of variation of between 45% and 58% (Keating et al. 1992). Temperatures vary from 14 °C to 29 °C with February and September being the hottest months and July and August the coolest months.

Maize is the most important cereal and is commonly intercropped with beans, cowpea, and pigeon pea. Coffee is an important cash crop in the Upper Midlands AEZ. The dairy cattle are mainly kept under stall feeding where feed is delivered. Improved forages such as Napier and Rhodes grass are planted but occupy only in a small proportion on the farm. Seasonal shortfall of feed resources is common due to prolonged drought and erratic rainfall and poor management of pastures.

Brachiaria Grass Validation Trial

The *Brachiaria* grass cultivars tested in this study were identified from an evaluation conducted on several accessions and cultivars in Kenya (Njarui et al. 2016). They were found to be productive and had positive responses to livestock productivity. The objective of the validation trial was to evaluate the growth and productivity of best bet *Brachiaria* grass cultivars. The validation trials were managed by researchers and were conducted at Kamweti ATC involving farmers.

Treatments, Field Design, and Management

Four best bet *Brachiaria* grass cultivars, namely, *Brachiaria decumbens* cv. Basilisk, *B. brizantha* cvs. MG-4, Piata, and Xaraes, were compared with Rhodes grass – a commonly cultivated grass in the region. The treatments were arranged in a randomized block design with three replications in plot sizes of 4 m × 5 m with 1 m alley between plots. Ten rows were planted in each plot at interrows spacing of 50 cm. The land was ploughed and harrowed to a fine tilth before planting the seeds. The seeds were drilled by hand along shallow furrows of about 0.5–1 cm deep at a rate of 5 kg/ha for all the grasses and covered lightly with soil. Triple Super Phosphate fertilizer (TSP 46% P_2O_5) was applied in the planting furrows prior to sowing the seeds at a rate of 40 kg/ha P.

The trial was established in March 2018, during the long rains (LR) season (March–May). A standardization cut was made after the end of first wet season in June 2018 in all plots to stimulate uniform plant growth. Calcium ammonium nitrate (CAN, 26% N) was applied at a rate of 50 kg N/ha/season and commenced after the standardization cut. The fertilizer was broadcasted in the plots and covered slightly using hand hoes, approximately 1 week after onset of rainy season. The plots were kept weed-free by weeding using handheld hoes until the grass were fully established.

Data Collection and Statistical Analysis

Weather and Agronomy Data

Rainfall data was recorded from the weather station near the experimental plots. The first wet seasons (LR 2018) was regarded as the establishment phase and subsequent seasons as production phase. During the production phase, the data collected included plant numbers, tiller numbers, plant height, pests and diseases damage, and dry matter (DM) yield at 8 weeks interval. Numbers of plants were determined by counting plants within a 1 m^2 frame, randomly placed over two central rows. Tiller numbers and plant height were recorded for four plants in the same quadrat. Plant height was measured for from ground to the tip of plant. Pests and diseases damage were rated on a 1–5 scale; where 0 = no damage and 5 = highest damage. For dry matter (DM) yield determination, an area of 2 m × 2 m fixed quadrat was sampled by cutting the plants to around 5 cm stubble height using handheld sickles. The fresh materials were weighed, a subsample taken where necessary, and dried at 105 °C for 48 h. The rainfall at Kamweti ATC from January 2018 to December 2019 is given in Fig. 2. The total monthly rainfall in March and April 2018 was greater than 2019 rainfall recorded during the same months, whereas October and December rainfall was higher in 2019 than in 2018.

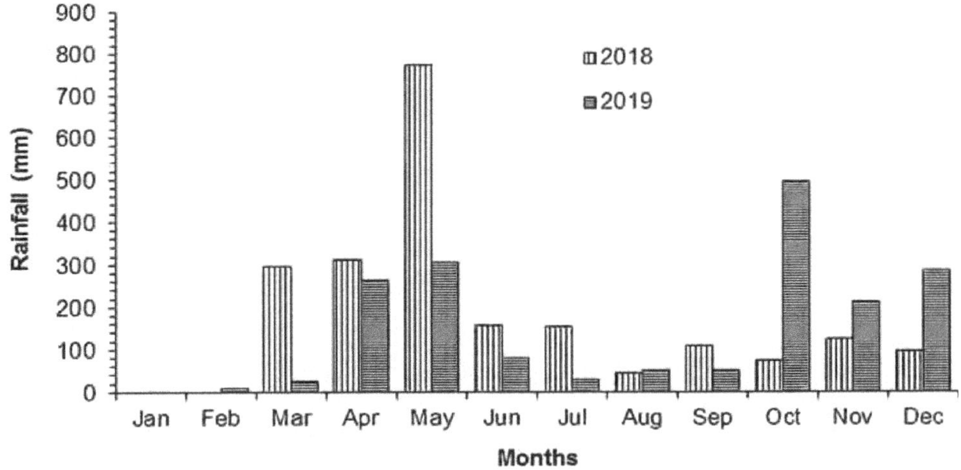

Fig. 2 Monthly rainfall during the experimental period, March 2018–December 2019 at Kamweti ATC, Kenya

Forage Quality Analysis

The herbage was analyzed for phosphorus, calcium, crude protein (CP), fibers, lignin, and ash for samples harvested in July 2019. The samples were dried at 65 °C in an oven to a constant weight and then ground with a hammer mill fitted with a 1 mm sieve, and about 200 g was preserved for analysis. Ash was determined by heating the samples at 550 °C for 8 h in a muffle furnace. Crude protein was determined using Micro-Kjeldahl according to the method of Association of Official Analytical Chemists (AOAC) (1990). The neutral detergent fiber (NDF), acid detergent fiber (ADF), and acid detergent lignin (ADL) were determined using the Ankom method of Van Soest et al. (1991).

Soil Mineral Analysis

The soil samples were taken from a depth of 0 to 30 cm in each grass plot and air-dried. Analysis was carried out for organic carbon (OC), total N, K available P, and pH. Soil pH was measured in water (soil/water ratio of 1:2.5).The N was determined following the Micro-Kjeldahl method after Bremner and Keeney (1965), while P was measured by ascorbic acid method (Watanabe and Olsen 1965). The total OC and K were measured according to methods described by Okalebo et al. (2002).

Participatory Evaluation of *Brachiaria* Grass Cultivars

The criterion for evaluation of the *Brachiaria* grasses was developed through a focus group discussion of 26 farmers randomly selected from the region. Scientists and extension officers guided the farmers. A pairwise ranking matrix was used to rank the listed plant attributes in their order of importance which were then applied in the *Brachiaria* evaluation process. At the end of wet season, 72 farmers (26 females, 46 males) participated in the evaluation of the *Brachiaria* grass cultivars and Rhodes grass in the validation trial established at Kamweti ATC. The field assessment was based on grass phenotypic traits (visual herbage production, tillering capacity, height, hairiness, and easy to cut). For each trait, farmers recorded their individual scores for each individual grass cultivars in an evaluation form using a Likert scale of 1–4, where, 1 = poor, 2 = average, 3 = good and 4 = very good.

On-Farm Evaluation of *Brachiaria* Grass Cultivars

Farmers who had previously participated in the project household baseline survey in Kangundo subcounty and Kirinyaga County were selected for *Brachiaria* grass on-farm trials. The criteria for selection of farmers were based on ownership of dairy cattle, willingness to participate in data collection, and ability of farmers to establish at least 0.1 ha of *Brachiaria* grass. The farmers were trained on establishment, management, and harvesting of the grasses. Individual farmers were provided with seeds of two cultivars (either Piata, Xaraes, or Basilisk). A total of 84 and 74 farmers received *Brachiaria* seeds in Kangundo subcounty and Kirinyaga County, respectively. Only 16 farmers were monitored in each site, and a farmer was considered as a replication. The land was prepared using local practices and sown in March and April 2018. Close supervision and monitoring was undertaken through joint action with farmers, extension, and researchers. A range of data was collected for assessment of ecological, food and nutrition security, and social-economic impact. For the ecological impact, the data collected includes *Brachiaria* biomass yield, soil organic carbon, effect of *Brachiaria* on milk production, and acreage under *Brachiaria*, while for the food and nutrition security impact, the data include availability of milk, level of consumption of milk, and dietary diversity. For the social-economic impact, the data collected include people aware of the technology, people growing *Brachiaria*, income from *Brachiaria*, and milk sales.

Strategies Used to Upscale *Brachiaria* Grass Cultivars

Different extension and advisory services (EAS) have been used to upscale the suitable *Brachiaria* grass cultivars in Kangundo and Kirinyaga counties and other areas in Kenya. Some of the EASs include VKC, field days, agricultural shows, posters, and linkages with other institutions with support of multi-actor platform. VKC is a digital-based system linking farming communities with shared interests through smart phones and social media platforms such as WhatsApp. The VKC has been adopted to disseminate technology. The process involved sensitizing farmers on benefits of the grass, providing information on planting, management, and harvesting followed by seed distribution. Each farmer was provided with seeds of two *Brachiaria* cultivars (40 g of each cultivar) enough to cover an area of 160 m^2.

Statistical Data Analysis

From the validation trial only data on plant tillers, height and dry matter yield were analyzed at each harvest. Likewise the values for forage quality were also statistically evaluated. The participatory evaluation scores from each farmer were entered in Microsoft Excel spreadsheet and a mean score for each criterion calculated for each grass cultivar by gender. The data were subjected to Analysis of Variance (ANOVA) using the statistical software GenStat 15 for windows (VSN International Ltd 2013). And where significant differences occurred, the means were separated using the Fisher's protected least significant difference (LSD) test at $p < 0.05$.

Results and Discussion

Brachiaria Grass Validation Trial

(i) *Plant Height*

Table 1 shows the mean plant height of four *Brachiaria* grass cultivars and Rhodes grass. Plant height at each harvest period varied among the grass cultivars and showed significant differences ($p < 0.05$). Rhodes grass was taller than all the *Brachiaria* cultivars in the first three harvests and in the

Table 1 Mean plant height (cm) of four *Brachiaria* grass cultivars and Rhodes grass (at 8 weeks cutting interval) at Kamweti ATC, Kenya

	2018		2019			
	Harvest 1	Harvest 2	Harvest 3	Harvest 4	Harvest 5	Harvest 6
Grass cultivars	29 Sept	20 Nov	15 Jan	12 Mar	4 July	22 Aug
Brachiaria decumbens cv. Basilisk	21.8^{B}	31.5^{B}	18.4^{B}	5.7^{A}	63.9^{B}	9.2^{A}
Brachiaria brizantha cv. MG-4	18.5^{B}	37.3^{B}	18.5^{B}	7.2^{A}	75.4^{B}	$8.9^{A,B}$
Brachiaria brizantha cv. Piata	13.5^{B}	43.0^{B}	11.0^{B}	7.0^{A}	27.9^{C}	6.4^{C}
Brachiaria brizantha cv. Xaraes	12.5^{B}	35.5^{B}	11.0^{B}	5.2^{A}	24.6^{C}	$7.2^{B,C}$
Rhodes grass cv. KAT R3 (control)	48.6^{A}	86.7^{A}	55.0^{A}	7.6^{A}	103.6^{A}	–[a]
LSD ($p < 0.05$)	13.1	12.3	14.2	N.S.[b]	17.3	1.9

Means with different superscript in the same column differ significantly at $P < 0.05$

[a]Rhodes grass dried therefore no plants were measured

[b]*N.S.* not significant

fifth harvest, but in the fourth harvest, there was no significant difference among grass cultivars tested (Table 1). Lack of significance difference among the grasses in the fourth harvest was mainly due to poor rainfall experienced during the growing period, which reduced the growth of all grasses. Similarly, the low plant height in sixth harvest was attributed to low rainfall and low temperatures in 2019 (see Fig. 2)

(ii) *Tiller Numbers*

All the *Brachiaria* cultivars had higher tiller numbers than Rhodes grass at all harvests (Table 2). Xaraes recorded the highest number of tillers in most of the harvests. Tiller numbers increased from less than 100 tillers/plant at first harvest to greater than 120 tillers/plant in the fourth harvest for all the *Brachiaria* cultivars and Rhodes and declined in subsequent harvest except for Basilisk which increased slightly in fifth harvest. *Brachiaria* forms a massive root development, and hence nutrient uptake is likely to be high allowing the plant to continue increase in tillers. Reduce tillers number in fifth and sixth harvest were attributed to low rainfall and temperatures (see Fig. 2). The higher tillering ability of *Brachiaria* grass than Rhodes grass is due to difference in growth habit where *Brachiaria* forms rhizomes, while Rhodes grass spreads by stolon. Moreover, number of tillers is an indicator of resource use efficiency and increases the chances of survival and influences herbage yield (Laidlaw 2005) if water and nutrients are not limiting.

(iii) *Dry Matter Yield*

Six harvests for the DM yield were conducted during the production phase (Fig. 3a–f). There were no significant differences ($p > 0.05$) on DM yield among the *Brachiaria* cultivars and Rhodes grass in the first three harvests and

Table 2 Mean number of tillers/plant of four *Brachiaria* grass cultivars and Rhodes grass (at 8 weeks cutting interval) at Kamweti ATC, Kenya

	2018		2019			
	Harvest 1	Harvest 2	Harvest 3	Harvest 4	Harvest 5	Harvest 6
Grass cultivars	29 Sept	20 Nov	15 Jan	12 Mar	4 July	22 Aug
Brachiaria decumbens cv. Basilisk	$83.3^{A,B}$	116.3^{A}	127.7^{A}	140.3^{B}	163.3^{A}	137.3^{A}
Brachiaria brizantha cv. MG-4	$89.3^{A,B}$	95.3^{A}	121.3^{A}	126.0^{B}	110.5^{B}	102.4^{A}
Brachiaria brizantha cv. Piata	$72.7^{B,C}$	78.7^{A}	142.7^{A}	151.7^{B}	$131.1^{A,B}$	120.8^{A}
Brachiaria brizantha cv. Xaraes	103.0^{A}	108.7^{A}	156.3^{A}	183.3^{A}	167.5^{A}	184.5^{A}
Rhodes grass cv. KAT R3 (control)	46.7^{C}	34.7^{B}	45.7^{B}	48.7^{C}	31.4^{C}	–[a]
LSD ($p < 0.05$)	30.3	43.5	55.3	29.9	46.5	N.S.[b]

Means with different superscript in the same column differ significantly at $P < 0.05$

[a]Rhodes grass dried therefore no plants were measured

[b]*N.S.* not significant

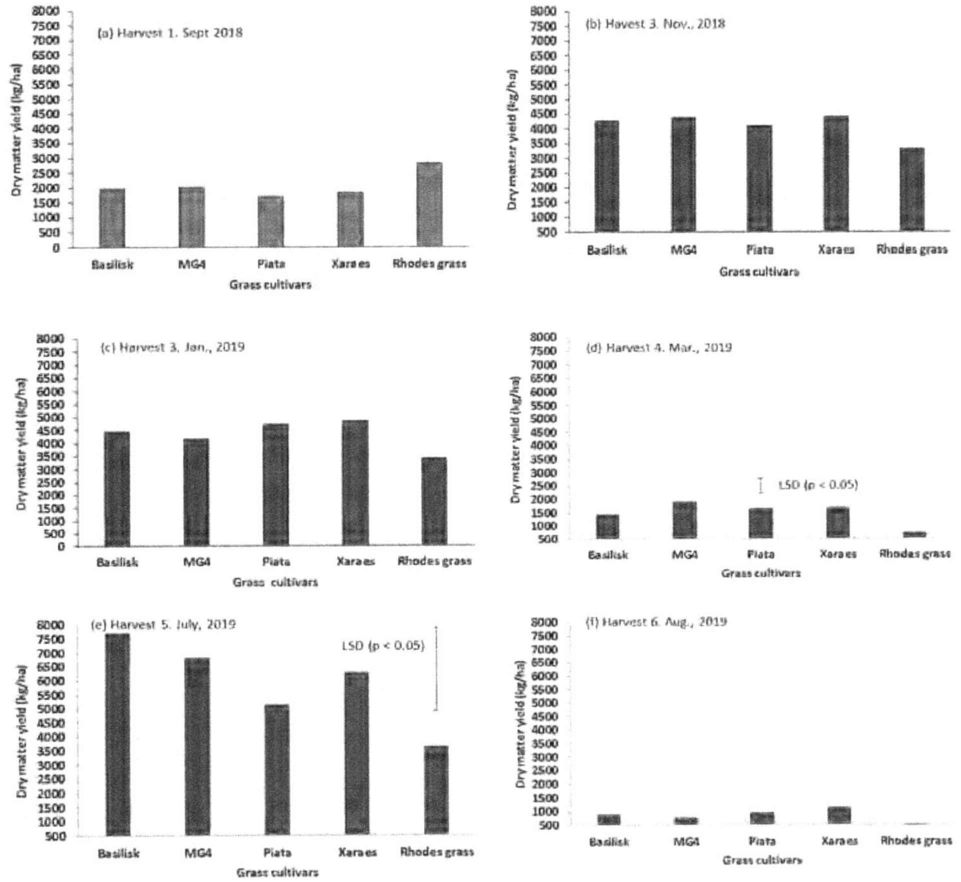

Fig. 3 Dry matter yield of four *Brachiaria* grass cultivars and Rhodes grass harvested every 8 weeks at Kamweti ATC, Kenya

sixth harvest. Significant ($p < 0.05$) difference in yield was recorded in the fourth (Fig. 3d) and fifth harvests (Fig. 3e). In the fourth harvest, all *Brachiaria* cultivars outyielded Rhodes grass, but yields were not significant different among themselves (Fig. 3). Similar studies by Nyambati et al. (2016) in an area that has unimodal rainfall pattern showed no significance difference among the same *Brachiaria* cultivars, but their yields were higher than those of Rhodes grass. However, in the fifth harvest, only Basilisk had higher ($P < 0.05$) DM yield (7.7 t/ha) than Rhodes grass (3.6 t/ha) but did not differ significantly with the other *Brachiaria* grass cultivars. Generally, *Brachiaria* yields were relatively higher in the second (4.1–4.4 t/ha), third (4.1–4.8 t/ha), and reached peak in the fifth (5.1–7.7 t/ha) harvests, and this was attributed to moisture availability. The first harvest (Fig. 3a) was conducted during the long dry season, while the fourth (Fig. 3d) was carried out toward the end of the short dry season. The second and third harvest was conducted during the wet season. The DM yields were appreciably higher than yield recorded by Njarui et al. (2016) in EM.

Table 3 Chemical composition (% of DM) of *Brachiaria* grass cultivars and Rhodes grass harvested at Kamweti ATC, Kenya

Grass cultivars	Calcium	Phosphorus	CP	ADF	NDF	ADL	Ash
Brachiaria decumbens cv. Basilisk	0.15^{A}	0.37^{A}	11.60^{A}	41.49^{B}	$67.21^{B,C}$	6.94^{B}	9.81^{B}
Brachiaria brizantha cv. MG-4	0.16^{A}	0.24^{A}	11.60^{A}	40.40^{B}	$67.37^{B,C}$	$7.05^{A,B}$	9.62^{B}
Brachiaria brizantha cv. Piata	0.13^{A}	0.53^{A}	11.73^{A}	36.82^{C}	$65.02^{C,D}$	6.00^{B}	9.54^{B}
Brachiaria brizantha cv. Xaraes	0.12^{A}	0.38^{A}	12.16^{A}	38.16^{C}	64.58^{D}	5.90^{B}	11.0^{A}
Rhodes grass cv. KAT R3 (control)	0.18^{A}	0.34^{A}	10.19^{A}	44.86^{A}	71.40^{A}	8.61^{A}	7.87^{C}
LSD ($P < 0.05$)	N.S.[a]	N.S.	N.S.	1.94	2.60	1.39	0.93
CV (%)	6.5	4.9	2.1	2.6	2.1	10.7	5.2

CP crude protein, *ADF* acid detergent fiber, *NDF* neutral detergent fiber, *ADL* acid detergent lignin
Means with different superscript in the same column differ significantly at $P < 0.05$
[a]*N.S.* not significant

(iv) *Forage Quality Composition*

The result of forage quality analysis is presented in Table 3. The calcium, phosphorus, and CP content were not significantly different among the tested grasses. Significant differences were recorded in ADF, NDF, ADL, and ash content. Rhodes grass had higher ADF and NDF than *Brachiaria* grass cultivars, while the ADL content did not differ from that of MG-4. Xaraes had the highest ash content (11%), and the Rhodes grass had the lowest. However, all *Brachiaria* cultivars contained more CP than those reported from the Eastern Midland (7–10%) by Njarui et al. (2016) but were lower than that reported from a similar region – CH (12–15%) by Nyambati et al. (2016). The lower CP content from this study was attributed to difference in the sampling time as the grasses were harvested during the dry season and growth was poor. Nevertheless, CP $\geq$ 11% during the dry season is an indication that *Brachiaria* can provide good quality feed in CH. All the *Brachiaria* met the minimum level of CP required for ruminant maintenance (7%) and milk production (11%). Moreover, the NDF was below 72%, above which generally decreases DM intake. All the *Brachiaria* cultivars had lower ash content (12–15%) than those reported by Njarui et al. (2016).

(v) *Soil Nutrients*

There was no statistical significance difference ($p > 0.05$) on soil properties (pH, N, org C, P, and K) among the tested *Brachiaria* grass cultivars and the Rhodes grass (Table 4). However, the total N (0.20–0.21%) and organic carbon contents (2.2–2.3%) in all *Brachiaria* cultivars were higher than the Rhodes

Table 4 Effects of *Brachiaria* grass cultivars and Rhodes grass on soil pH, organic carbon, and nutrients (nitrogen, phosphorus, and potassium), 18 months after planting

Grass cultivars	Soil pH (1:2.5 soil:H_2O)	Nitrogen (%)	Org. carbon (%)	Phosphorus (ppm)	Potassium (me %)
Brachiaria decumbens cv. Basilisk	4.84	0.21	2.28	20.0	0.09
Brachiaria brizantha cv. MG-4	4.89	0.21	2.34	18.3	0.11
Brachiaria brizantha cv. Piata	4.74	0.20	2.29	15.0	0.09
Brachiaria brizantha cv. Xaraes	5.01	0.21	2.33	21.7	0.09
Rhodes grass cv. KAT R3 (control)	4.91	0.16	1.74	20.0	0.11
LSD ($P < 0.05$)	N.S.[a]	N.S.	N.S.	N.S.	N.S.
CV (%)	3.6	13.9	15.4	22.3	15.9

[a]*N.S.* not significant

grass. The org C contents in the *Brachiaria* fields will improve the organic matter content of the soils, which in turn will enhance soil fertility and increase productivity. This makes *Brachiaria* grass suitable for adaptation and mitigation measures to climate change effects in Kenya and possibly in other sub-Saharan African countries. However, the effect of *Brachiaria* on soil pH would require relatively a longer period.

Brachiaria Grass On-Farm Trials

The mean DM yield of *Brachiaria* grass cultivar in Kirinyaga and Kangundo is given in Table 5. There were no statistical significant differences ($p > 0.05$) on DM yield among the *Brachiaria* cultivars grass within and between sites. Of the 16 farms monitored in each site, there were large variations in yield for each *Brachiaria* cultivars tested.

The large difference in yield between farms for a given cultivar was attributed to farmers' failure to follow recommended agronomic practices such as proper plant spacing, lack of weeding at establishment, limited use of fertilizer, etc. Generally, where the yields were poor the grasses were established in eroded and less fertile part of farm and no fertilizer application. In contrast in farms where the yields were high, the land was well terraced, and farmer had adopted proper soil conservation practices. Additionally, farmers had applied either cattle manure or inorganic fertilizer or both. The higher yield achieved in Kirinyaga was attributed to more rainfall experienced from June to October unlike in Kangundo where there was no rains resulting in the cessation of plant growth.

Table 5 Mean DM yield among the *Brachiaria* grass cultivars in Kirinyaga and Kangundo study sites

Brachiaria grass cultivars	Kirinyaga	Kangundo	LSD ($p < 0.05$)
	DM yield (kg/ha)		
Brachiaria decumbens cv. Basilisk	6702 (1704–18,156)	5485 (1081–12,204)	N.S.
Brachiaria brizantha cv. Piata	6884 (2548–15,363)	6447 (3510–10,922)	N.S.
Brachiaria brizantha cv. Xaraes	5108 (1733–12,962)	5680 (1175–12,856)	N.S.
LSD ($p < 0.05$)	N.S.[a]	N.S.	

Values in parenthesis are range of DM yield
[a]*N.S.* not significant

Box 1

Effects of* Brachiaria *forage on social-economic conditions: **By August 2019 approximately 1320 smallholder farmers (18.2% females) mainly involved in dairy cattle farming had been provided with *Brachiaria* grass seeds in Kirinyaga County. In Kangundo, 720 smallholder farmers (23% were females) received the seeds. Report from field indicates that each farmer has given out splits to at least two other farmers. Conservative estimate put the numbers of farmers to over 4000 farmers implying a high rate of uptake of the technology in the project areas. Seeds were also distributed to 1200 farmers in Nyeri County and over 300 farmers in other parts of Kenya. So far, the project has not found any farmer who has abandoned growing *Brachiaria*. Farmers are enthusiastic about the grass, and a good number have expanded the acreage under *Brachiaria* grass. Farmers have also reported increase in milk production from dairy cattle fed *Brachiaria* grasses. However, there were variation in terms increase, but the values were within the range of 15–40% recorded by Muinga et al. (2016). This implies *Brachiaria* has contributed to increased milk availability for consumption at household level and for sale to generate income. However, there are plan to carry out on-farm feeding trial jointly with farmers to ascertain the actual benefits of *Brachiaria* in term of milk production. In Kirinyaga, youths have formed groups with aim of baling *Brachiaria* hay for sale to generate income. (Source: Field Survey)**

Participatory Selection of *Brachiaria* Grass Cultivars

During the FGD, farmers identified 16 plant attributes that are important when selecting grasses (Table 6). Grasses that give high milk yield when fed to livestock were ranked first followed by highly nutritious grasses while color was the least important criteria. These findings were consistent with earlier work carried

Table 6 Pairwise ranking matrix of farmers' criteria in selecting grasses for livestock production in Kirinyaga, Kenya

Plant attributes	FG	T	DT	HT	DR	PA	PR	HMY	HNV	LH	DGC	EC	PE	EE	EM	HBY	Total	Rank
Fast growth (FG)		T	DT	HT	DR	PA	PR	HMY	HNV	LH	FG	EC	PE	EE	EM	HBY	1	15
Tall (T)			DT	HT	DR	PA	PR	HMY	HNV	LH	DGC	EC	PE	EE	EM	HBY	1	14
Drought tolerant (DT)				HT	DR	PA	PR	HMY	HNV	LH	DT	EC	PE	EE	EM	HBY	3	13
High tillering (HT)					DR	PA	PR	HMY	HNV	LH	HT	EC	PE	EE	EM	HBY	4	12
Disease resistant (DR)						PA	DR	HMY	HNV	DR	DR	DR	DR	DR	DR	DR	12	3rd
Highly palatable (PA)							PA	HMY	HNV	LH	PA	EC	PE	EE	EM	HBY	7	9
Pest resistance (PR)								HMY	HNV	PR	PR	PR	PR	PR	PR	PR	11	5
Give high milk yield (HMY)									HMY	HMY	HMY	HMY	HMY	HMY	HMY	HMY	15	1st
High nutritive value (HNV)										HNV	HNV	HNV	HNV	HNV	HNV	HNV	14	2nd
Less hairy (LH)											LH	LH	PE	EE	EM	HBY	7	10
Dark green color (DGC)												EC	PE	EE	EM	HBY	0	16
Easy to cut (EC)													PE	EE	EM	HBY	6	11

(continued)

Table 6 (continued)

Plant attributes	FG	T	DT	HT	DR	PA	PR	HMY	HNV	LH	DGC	EC	PE	EE	EM	HBY	Total	Rank
High persistence (PE)														PE	PE	PE	11	4
Easy to establish (EE)															EM	HBY	8	8
Easy to manage (EM)																HBY	9	7
High biomass yield (HBY)																	10	6

elsewhere in Kangundo by Gatheru et al. (2016). They reported that forages that gave high milk yield when fed to cattle and nutritionally balanced were priority traits in selecting feed for livestock. There were significant differences ($p < 0.05$) in farmers' scores for herbage, tillers, and tallness between the cultivars (Table 7). Overall, Basilisk had the highest mean score for the five attributes followed by MG-4, while Piata had the lowest. There were no significant differences on farmers' scores for hairiness and ease of cutting between the cultivars. The ranking order for the cultivars was not different between females and male farmers (Fig. 4).

Table 7 Farmers' mean scores for selected criteria from evaluation of *Brachiaria* cultivars

	Selection criterion						
Grass cultivars	Herbage production	Tillering capacity	Height at harvest	Hairiness	Easy to cut	Mean score	Rank
Basilisk	3.40	3.36	3.23	2.66	3.28	3.19	1
MG-4	3.19	3.19	2.70	2.99	3.09	3.03	2
Piata	2.32	2.49	2.00	2.67	3.01	2.50	5
Xaraes	2.63	2.79	2.21	2.99	3.12	2.75	3
Rhodes grass	2.16	1.85	3.13	2.97	3.03	2.63	4
LSD ($p < 0.05$)	0.31	0.29	0.29	N.S.[a]	N.S.		

[a]*N.S.* not significant

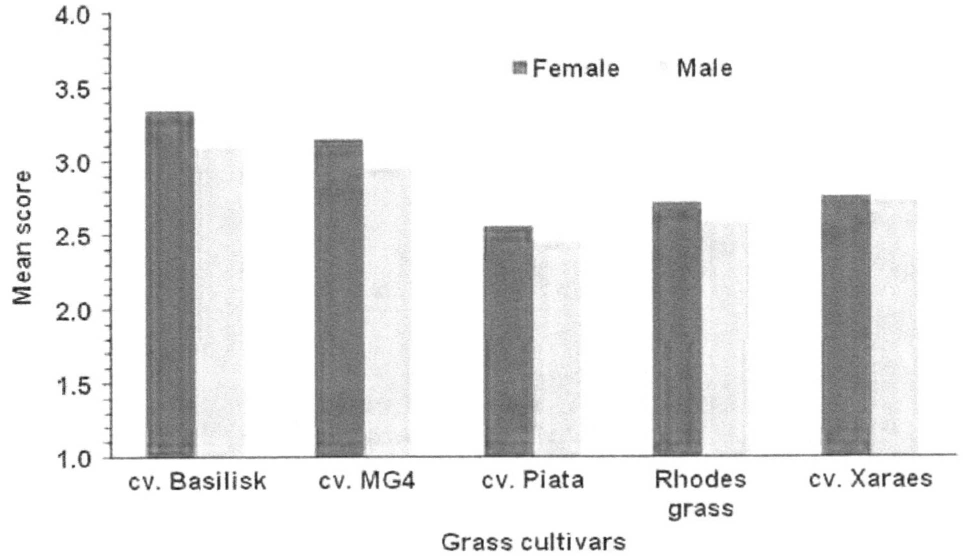

Fig. 4 Evaluation score for *Brachiaria* grass cultivars and Rhodes grass categorized by gender

Upscaling *Brachiaria* Grasses Cultivars

Multi-actor Platform: The multi-actor platform (MAP) members are composed from institution in private sectors, policy makers, dairy cooperative societies, extension services, farmers, and researchers. The MAP members were involved at InnovAfrica project inception phase during planning, site selection, and development of implementation framework. The MAPs have contributed in *Brachiaria* grass validation trial particularly in selection of farmers and development of selection criteria. They participate in field project evaluation and monitoring every 6 months and provide useful feedback to correct any divergence from the planned activities. They are also involved in upscaling of *Brachiaria* grass technologies through field days and distribution of seeds. For example, one MAP member who is committee of Mukurwe-ini Wakulima Dairy Cooperative Society has distributed seeds to over 1200 farmers who are members of the society in Nyeri County.

Village knowledge Centre: One VKC was established in Kangundo village of Kangundo subcounty in May 2018. From May 2018 to December 2019, a total of 530 people (comprising of farmers, extension agents the public sector, and non-governmental organization) and other stakeholders visited the VKC (Fig. 5) primarily to get information on *Brachiaria* grasses and on other issues (e.g., horticulture, poultry, bee, rabbits, microfinance). Most of the visitors to the VKC were over 50 years old. The VKC has connected over 300 farmers on social media (two WhatsApp Group). These have provided a platform where information on cultivation of *Brachiaria* grass including management, conservation, and feeding is shared. However, social networking is an important mean for promoting technology adoption (Brhane et al. 2019).

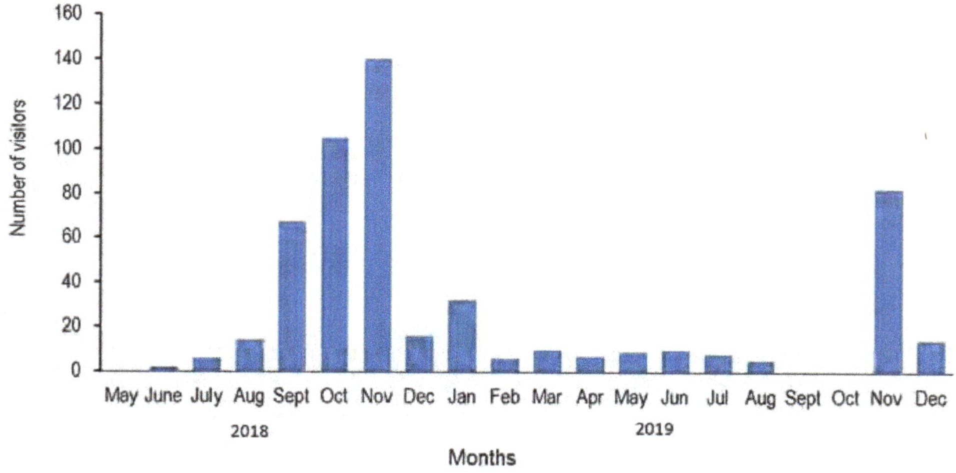

Fig. 5 Number of visitors to the Village Knowledge Centre from May 2018 to December 2019 in Kangundo, Kenya

Box 2 Other *Brachiaria* Upscaling Strategies Adopted

Field days: In May 2018, a farmer field day was held in Kangundo village to provide first-hand information to farmers and relevant stakeholders on *Brachiaria*-livestock value chain and related technologies and its impact on the feed security and livestock productivity. A total of 93 people representing smallholder dairy farmers, members of a small-scale dairy processing group, County extension agents, local leaders, and InnovAfrica partners attended.

Agricultural show and exhibition: InnovAfrica project participated in Agricultural Show of Kenya (ASK), Machakos Branch, in June 2018 and 2019 to create awareness on the *Brachiaria* technologies. The grasses were planted in small plots within the ASK ground for exhibition to farmers. Posters with details on the establishment, management, and nutritive quality of the grasses were also displayed. In 2018 an estimated 770 show attendants (farmers, scholars, and students) visited the *Brachiaria* grass plots, and the number increased to 1050 in 2019.

Posters: The project has also adopted posters to create farmer awareness on the availability of *Brachiaria* grass seeds for planting. Over 550 and 770 posters written in vernacular were distributed in Kirinyaga County and Kangundo subcounty, respectively. The posters were displayed at strategic locations where they were easily visible. This included trees along rural roads, in market centers, petrol stations, churches, dairy cooperatives, premises, and other public buildings where the local people visit frequently.

Agricultural Training Centre: The project has partnered with Kamweti ATC and Wambugu ATC in Kirinyaga and Nyeri Counties, respectively, to accelerate the upscaling of *Brachiaria* grass technology. These Centers normally organize field days, trains farmers, and conduct demonstrations on proven technologies on crops, forages, and livestock. The role of the ATCs is to bulk the *Brachiaria* grass and distribute splits to farmers.

Conclusions and Way Forward

This study is a farmer led experimentation to test *Brachiaria* grass for livestock production in Kirinyaga County and upscaling it in Kangundo subcounty. The validation trial showed that all the *Brachiaria* grass cultivars were more productive than the control (Rhodes grass) in most of the harvests. Similarly, based on phenotypic traits used by farmers to evaluate the grass, all the *Brachiaria* grass cultivars had higher mean score than Rhodes grass. Thus, *Brachiaria* has a potential to improve the feed availability in the region tested and other areas of Kenya and enhance food and nutrition security. The org C contents in the *Brachiaria* fields will improve the organic matter content of the soils which in turn will enhance soil fertility and increase productivity. This makes *Brachiaria* grass suitable for use in adaptation and mitigation measures to climate change effects in Kenya and possibly in other SSA countries. Over 4000 farmers are now engaged in cultivation *Brachiaria* grass in the study sites and additional 1500 farmers in other parts of Kenya, because of increased awareness by the InnovAfrica project using innovative EASs including VKC, ASK, posters, social media, and MAPs engagement. The demand for *Brachiaria* seeds is increasing due to benefits gained in terms of increased milk from livestock fed on the grass. Nevertheless, there is need to continue to upscale *Brachiaria* grass using already existing avenues and/or establish new ones.

Acknowledgements This is funded by the European's Union H2020, research and innovation program under Grant Agreement No. 727201.

References

Association of Official Analytical Chemists (AOAC) (1990) Official methods of analysis of the AOAC, 15th edn. AOAC, Arlington, pp 1–1298

Bremner JM, Keeney DR (1965) Steam distillation methods for determination of ammonium, nitrate and nitrite. Anal Chem Acta 32:485–495

Brhane G, Fantaye Belay F, Gebreselassie T, Desta D (2019) Enhancing sorghum yield through demonstration of improved sorghum varieties in Tanqua-Abergelle Wereda, Central Zone of Tigray, Ethiopia. J Agric Ext Rural Dev 11:11–16. https://doi.org/10.5897/JAERD2018.1020

Christensen NJH, Hewitson B, Busuioc A, Chen A, Gao X, Held I, Jones R, Kolli RK, Kwon W-T, Laprise R, Magana Rueda V, Mearn SL, Menendez CG, Raisanen J, Rinke A, Sarr A, Whetton P (2007) Chapter 11: Regional climate projections. In: Climate change 2007: the physical science basis. Contribution of working group I to the Fourth Assessment Report of the Intergovernmental Panel on Climate Change. Cambridge University Press, Cambridge, pp 847–940

Djikeng A, Rao IM, Njarui D, Mutimura M, Caradus J, Ghimire SR, Johnson L, Cardoso JA, Ahonsi M, Kelemu S (2014) Climate-smart Brachiaria grasses for improving livestock production in East Africa. Trop Grassl-Forrajes Trop 2(1):38–39

Freier KP, Bruggemann R, Scheffran J, Finckh M, Schneider UA (2012) Assessing the predictability of future livelihood strategies of pastoralists in semi-arid Morocco under climate change. Technol Forecast Soc Chang 79(2):371–382

Gatheru M, Njarui DMG, Gichangi EM (2016) Participatory evaluation and selection of suitable *Brachiaria* grass cultivars for production in the semi-arid eastern Kenya. In: Njarui DMG, Gichangi EM, Ghimire SR, Muinga RW (eds) Climate smart *Brachiaria* grasses for improving livestock production in East Africa – Kenya experience. Proceedings of the workshop held in Naivasha, Kenya, 14–15 Sept 2016. Nairobi, pp 124–131

Ghimire S, Njarui D, Mutimura M, Cardoso J, Johnson L, Gichangi E, Teasdale S, Odokonyero K, Caradus J, Rao I, Djikeng A (2015) Climate-smart *Brachiaria* for improving livestock production in East Africa: emerging opportunities. In: Vijaya D, Srivastava M, Gupta C, Malaviya D, Roy M, Mahanta S, Singh J, Maity A, Ghos P (eds) Sustainable use of grassland resources for forage production, biodiversity and environmental protection. Proceedings of 23rd International Grassland Congress, 20–24 Nov 2015, New Delhi, pp 361–370

Jaetzold R, Schmidt H, Hornetz B, Shisanya C (2006) Farm management handbook of Kenya, vol. II: natural conditions and farm management information. Part C: East Kenya Subpart C2 Coast Province, 2nd edn. Ministry of Agriculture and GIZ, Nairobi

Keating BA, Siambi MN, Wafula BM (1992) The impact of climatic variability on cropping research in semi-arid Kenya between 1955 and 1985. In: Probert ME (ed) A search for strategies for sustainable dryland cropping in semi-arid Eastern Kenya. ACIAR proceedings no. 41, Canberra, pp 16–25

Kenya National Bureau of Statistics (KNBS) (2010) The 2009 Kenya population and housing census: vol. II: population and household distribution by socio-economic characteristics. Kenya National Bureau of Statistics, Nairobi, p 438

Kusewa P, Guiragossion V (1989) Research priorities for enhancing crop productivity in marginal areas of Kenya. In: Agricultural research in Kenya, achievements, challenges and prospects. Proceedings of the 1st KARI scientific conference held at Panafric Hotel Kenya, 14–16 Aug 1989, pp 21–36

Laidlaw A (2005) The relationship between tiller appearance in spring and contribution of dry-matter yield in perennial ryegrass (*Lolium perenne* L.) cultivars differing in heading date. Grass Forage Sci 60:200–209

Lusweti CM, Nandasaba J, Onginjo E, Asena D (2004) Preliminary results of disease survey on Napier grass in selected sites of Western Kenya. Pasture research annual report 2004. National Agricultural Research Centre, Kitale

Martin R, Linstädter A, Frank K, Müller B (2016) Livelihood security in face of drought – assessing the vulnerability of pastoral households. Environ Model Softw 75:414–423

Muinga RW, Njunie MN, Gatheru M, Njarui DMG (2016) The effects of *Brachiaria* grass cultivars on lactation performance of dairy cattle in Kenya. In: Njarui DMG, Gichangi EM, Ghimire SR, Muinga RW (eds) Climate smart *Brachiaria* grasses for improving livestock production in East Africa – Kenya experience. Proceedings of the workshop held in Naivasha, Kenya, 14–15 Sept 2016. Nairobi, pp 229–237

Ngila P, Njarui DM, Musimba NK, Njunie MN (2016) Performance of Galla goats fed different cultivars of *Brachiaria* in the coastal lowlands of Kenya. J Fish Livest Prod 5:210. https://doi.org/10.4172/2332-2608.1000210

Njarui DMG, Gichangi EM, Ghimire SR, Muinga RW (eds) (2016) Climate smart *Brachiaria* grass for improving livestock production in East Africa – Kenya experiences. Kenya Agricultural and Livestock Research Organization, Nairobi, p 271. https://cgspace.cgiar.org/handle/10568/79797

Nyambati EM, Ayako W, Mailu S, Chelimo EJ, Njarui DMG (2016) Production and nutritive quality of Brachiaria grass cultivars subjected to different cutting intervals in the cool sub-humid highlands of central Kenya. In: Njarui DMG, Gichangi EM, Ghimire SR, Muinga RW (eds) Climate smart *Brachiaria* grasses for improving livestock production in East Africa – Kenya experience. Proceedings of the workshop held in Naivasha, Kenya, 14–15 Sept 2016. Nairobi, pp 62–69

Okalebo JR, Gathua KW, Woomer PL (2002) Laboratory methods of soil and plant analysis: a working manual, 2nd edn. TSBF-CIAT and SACRED Africa, Nairobi, p 128

RoK (Republic of Kenya) (2019) Ministry of Agriculture, Livestock, Fisheries and Irrigation, State Department for livestock. Draft National Livestock Policy, Kenya, p 75

Schilling J, Freier KP, Hertig E, Scheffran J (2012) Climate change, vulnerability and adaptation in North Africa with focus on Morocco. Agric Ecosyst Environ 156:12–26

Simpson JR, Okalebo JR, Lubulwa G (1996) The problem of maintaining soil fertility in Eastern Kenya. A review of relevant research. ACIAR monograph no. 41, Canberra, p 60

Thornton PK, van de Steeg J, Notenbaert A, Herrero M (2009) The impacts of climate change on livestock and livestock systems in developing countries: a review of what we know and what we need to know. Agric Syst 101:113–127

Van Soest PJ, Robertson JD, Lewis BA (1991) Methods for dietary fibre, neutral detergent fibre and non-starch polysaccharides in relation to animal nutrition. J Dairy Sci 74:3583–3597

VSN International Ltd (2013) Genstat for windows, 15th edn. VSN International Ltd, Hemel Hempstead

Watanabe FS, Olsen SR (1965) Test of an ascorbic acid method for determining phosphorus in water and sodium bicarbonate extracts from soils. Soil Sci Soc Am J 29:677–678

World Bank (2013) Turn down the heat: climate extremes, regional impacts, and the case for resilience – a report for the World Bank by the Potsdam Institute for Climate Impact Research and Climate Analytics. World Bank, Washington, DC, p 254

9

Green Technology Approaches to Solid Waste Management in the Developing Economies

T. B. Hammed and M. K. C. Sridhar

Contents

Abstract

The severity of extreme weather and climate change impacts around the world has been a public health concern in the last few decades. Apart from greenhouse gas generation, poor waste management exacerbates consequences of global warming such as flooding, lower crop yields, and the epidemic of diseases which can escalate into disastrous situations. The general public in developing economies sees wastes as valueless materials and disposes them through open burning, stream dumping, or as conveniently as possible. Also, the cutting of trees for firewood leads to deforestation and desertification that increase people's vulnerability to climate change impact. Against this backdrop, there is a need for a paradigm shift toward developing indigenous technologies that convert solid waste to cheap and clean energy. Various innovations use the "green technology approach" in putting trash back into the value chain. Furthermore, the green technology approach has a great potential to enhance adaptation and resilience

T. B. Hammed (✉) · M. K. C. Sridhar
Department of Environmental Health Sciences, Faculty of Public Health, College of Medicine, University of Ibadan, Ibadan, Nigeria

among climate change-displaced populations where they can set up microenterprise on useful end products. In this chapter, unique features of these technologies at the Renewable Resources Centre of the University of Ibadan, practice-oriented researches, and a case study at Kube-Atenda community Ibadan, Nigeria, are presented. This chapter is therefore set out to showcase examples of waste management initiatives and strategies that have been successfully implemented elsewhere by the authors. It also focuses on how some countries in the continent, with developing economies, may foster their resilience and their capacity to adapt to climate change.

Keywords

Climate change · Waste management · Greenhouse gas · Indigenous technologies · Gaseous emission · Value chain

Introduction

Africa continent is at the forefront of vulnerability to global climate change within the world. The overall public in this continent sees wastes as valueless materials and disposes them through open burning, stream dumping, or as conveniently as possible (Hammed et al. 2018). Healthcare wastes are burnt in low-temperature incinerators with gaseous emissions. Also, the cutting of trees for firewood ends up in deforestation and desertification that increase people's vulnerability to temperature change impact. As an example, many households in Nigeria (up to 70%), including households in rural areas (86%), and households in urban areas (42%), depend on solid fuel as their primary source of energy (Sridhar et al. 2016). The majority of the families that use solid fuel for cooking (94%) operate open fire or stoves that have no chimneys or hood (Gwatkin et al. 2000). Other than climate change, the utilization of firewood as fuel for cooking exposes people, and most children and women, to smoke which poses a health risk to them and might cause respiratory illnesses (Salvi and Barnes 2010; Oluwole et al. 2012). Consistent with some researchers (Smith et al. 2001; Dehoust et al. 2005), the waste management sector makes a relatively minor contribution to greenhouse gas (GHG) emissions at a global scale, estimated at approximately 3–5% of total anthropogenic emissions in 2005. Also, the average annual per capita waste generation in developing nations is estimated at 10–20% of that of developed countries, which is rising in response to economic growth. The waste sector is, however, in an exceedingly unique position to maneuver from being a secondary source of world emissions to becoming a major saver of emissions. It is on this premise that the Chief Director of the United Nations Environment Programme (UNEP) has been developing a technique for cobenefits of waste management within the context of global climate change. Integrated Solid Waste Management (ISWM) based on the 4Rs (reduce, recycle, reuse, and recovery) approach, with a special focus on sustainable consumption and production, E-waste management, conversion of agricultural biomass and waste plastics into energy and material resources, and management of hazardous waste, is being considered by the UNEP (UNEP 2010).

A recent report by Oyelaran-Oyeyinka of the UN Human Settlements Programme (UN-Habitat) (The Guardian, 27 April 2015) indicated that "African cities are the engines of wealth creation and therefore the centers of manufacturing, production, and innovation." This fact has not yet been justified and, as such, many cities have changed to slums that are characterized with incessant poverty, homelessness, the prevalence of preventable diseases, environmental pollution, maternal and child mortality, and other problems. These problems are further aggravated by social, economic, and environmental issues posed by climate change. There is, therefore, a need to strategize how African nations may foster their resilience and adaptation to the implication of climate change. Moreover, UNEP has identified two categories of solid waste management techniques that can reduce GHG emissions or contribute a net benefit to the development (UNEP 2010). The first set of technologies in view is technologically simple, locally available, and relatively cheap. The other set involves sophisticated, expensive technology and typically requires over 5 years to implement due to some factors, including financing, planning, and regulatory processes. The first process is seen to be very useful to strengthen resilience and adaptive capacities to counteract climate-related hazards within the developing economies. This chapter showcases samples of successful waste management initiatives and strategies that have been successfully implemented at the grassroots to build the capacity of the rural and urban dwellers, and foster their resilience to the menace of climate change.

Green Technology Approach to Waste Management and Climate Change Mitigation and Resilience in the Developing Economies

The significant problems of solid waste management in low-economic settings include lack of interaction and collaboration between stakeholders (community members, researchers, industrialists, and policymakers) in the policies that deal with waste. Also, there is a lack of political will for a waste-recycling program, no provision for research funds, unavailability of sustainable technologies and over-dependence on imported technology, and lack of transparency and marketplace for recycled products. There is usually no sense of belonging in utilizing any project in which the end-users are not involved and motivated (McDonald and Ball 1998; Tucker et al. 1998; Paula and Elizabeth 2008; Fauzul and Viradin 2011; Hammed et al. 2012). Consequently, waste disposal systems do not conform to best practices, and in most cases, there is no sound or acceptable disposal method for solid waste in place. These wastes are usually heaped in piles within residential areas and oftentimes burnt once they overflow, or the stench from them becomes unbearable, creating air pollution-associated issues. The disposal and treatment of wastes lead to GHG emissions with severe global climate change impacts. Methane is the most significant GHG gas emanated from solid waste in the landfills, which is released during the breakdown of deposited organic matter (IPCC 2006).

The gross of socioeconomic, environmental, and health difficulties the climate change phenomenon brings to the human being is depletion of natural resources, atmospheric pollution, outbreaks of vector-borne diseases, and flooding as well as crop failures that lead to malnutrition. Ironically, these impacts are mostly strongly seen and felt in developing countries, whose contributions to GHG emissions are far lower than their counterparts in the industrialized world. The aforementioned has called for effective initiatives and strategies such as those that supply renewable power from waste, mitigate climate change, create job opportunities, and promote growth in Gross Domestic Product (GDP) through wealth generation mechanisms. Generally, the benefits of green technology approach in the form of 5Rs (reduce, reuse, recycle, recover, and repair) to waste management include:

- Conservation of natural resources
- Environmental sanitation and protection from soil erosion, gully problems, deforestation, and desertification as trees are not cut
- Prevention of biodiversity loss
- Reduction of demand for landfilling of wastes
- Job creation and poverty eradication in the state; create wealth among peasant farmers
- Food security in the country through organic fertilizer production
- Promotion of local technology for sustainable development
- Reduce the amount of CO_2 and methane gas being released into the atmosphere
- Saves foreign exchange for the nation

Considering the hierarchy of sustainable waste management, waste prevention is generally far better than any other waste management practice. That is, waste prevention eliminates greenhouse emissions and other environmental problems that are resulted from waste disposal and also ensures fewer resource extractions and manufacturing (the United States Environmental Protection Agency 2006). Following waste prevention, in terms of benefit, is waste recycling which may be a closed-loop or open-loop process. The open-loop recycling process occurs when recycled material is employed to form a completely new item. During this case, there is a loss of fabric quality which is brought up as "down-cycling." Another strategy being employed in line with waste recycling is industrial symbiosis. Here, industries exchange useful by-products among themselves (Ashton et al. 2009). Sometimes, the industries involved in this arrangement may form "recycling clusters" to facilitate the sharing of resources (Chertow and Lombardi 2005; Harris 1999). A good example is the United kingdom's National Industrial Symbiosis Programme that has successfully saved more than five million metric tons of wastes that could have ended in landfills and eliminated more than five million metric tons of GHG emissions since 2005 (Chertow 2009). The above state of affairs illustrates the necessity for identification of processes, methods, and tools which may help African nations from international best practices. This approach calls for global thinking and local action.

Field-Oriented Activities at Renewable Resource Centre, University of Ibadan

The Renewable Resource Centre (RRC), located in the poultry unit of the Teaching and Research Farm, University of Ibadan, Nigeria (Fig. 1), was established in the year 2013. The center started with a few viable technologies developed in the Department of Environmental Health Sciences of the University which were scaled up in the field at several locations nationwide. The techniques are now proven to be viable to exploit commercially. The RRC has several sections, which are shown in Fig. 2:

(i) A building with training facilities to train up to 30 trainees, a storeroom, a research exhibition area (TECH park) to display environmental health research outputs, a caretaker's restroom, two toilets for visitors, a well-equipped mushroom spawn production area, and a small lounge where a visitor can sit and relax.
(ii) Biogas digesters to generate biogas from poultry piggery and cattle wastes for lighting and compression into cylinders.
(iii) A mushroom cultivation hut with thatched roof and bamboos, and accessories to produce on a large scale.

Fig. 1 The RRC main building in the university

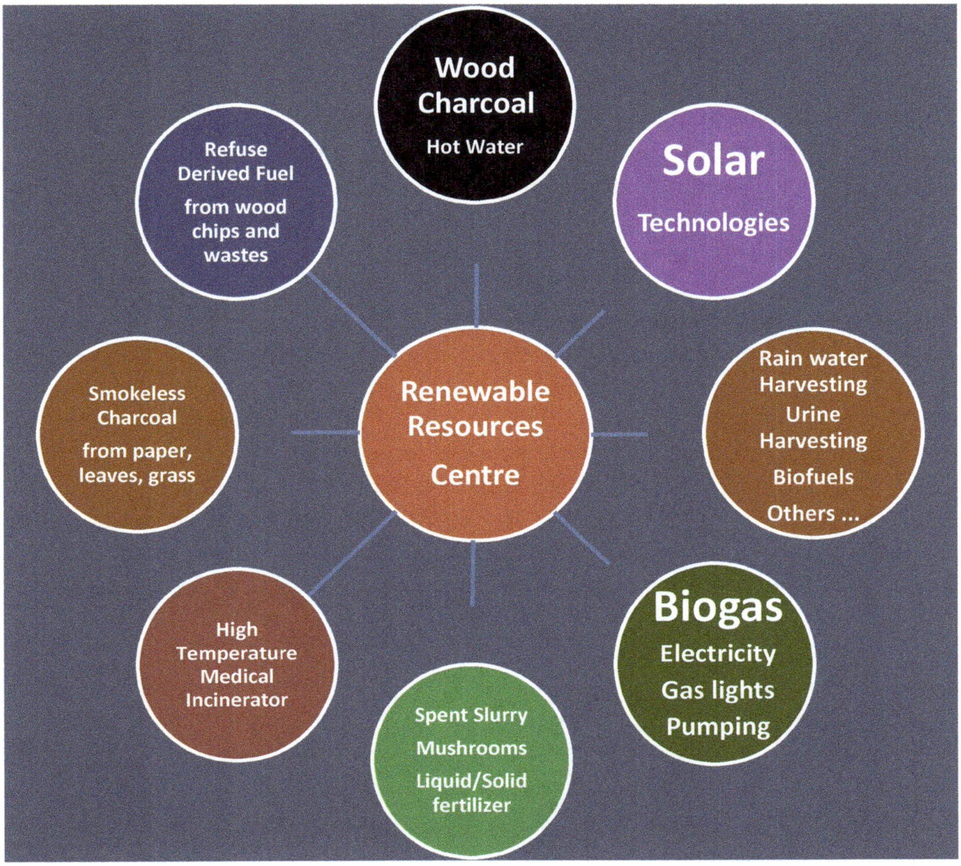

Fig. 2 Various sections at Renewable Resources Centre

(iv) Liquid fertilizer production from biogas-spent slurry.
(v) Permanent kiln for smokeless charcoal/biochar production.
(vi) Charcoal Kiln (a design adopted from Kenyan/Ethiopian technology) to produce charcoal within 24 h from fallen and cut trees.
(vii) Home composter for low-scale production of compost at house and farm unit.
(viii) Portable refuse-derived fuel unit.
(ix) A demonstration plot land set aside opposite the biogas section for demonstrating some of the products for agricultural production. Here, various field-based experiments are carried out.
(x) Rainwater harvesting for ensuring the availability of water all year round with elementary and useful technology.

It is expected that the RRC will become a center of attraction for people to get trained on "Waste to Wealth" and "Waste to Energy" technologies. The center will benefit Nigerian communities as well as other developing economies through "Town and Gown" meet.

Activities and Climate Change Adaptive Technologies at Various Sections of the Center

Training Hall: About 30 people can comfortably sit and receive training. The room is equipped with two air conditioners, a PowerPoint projection, and a cupboard to display some standard books for the trainees to use. The target training participants include interested individuals, small and big scale farmers, members of cooperative society, students, and retirees (Fig. 3a).

Biogas Generation and Compression

There are two types of biogas digesters: One is 8 m^3 capacity, flexible, and a product of PVC which can expand as the gas is generated (Fig. 4a). This kind of digester is suitable for larger families or establishments that need more gas for consumption. The other is a unique floating dome variety; the capacity can vary from 1 to 10 m^3 or more, built from readily available water storage tanks from the local markets. They are affordable for households and have the potential for promotion at community levels to individual families. The biogas plants are designed to use dairy waste, poultry waste, piggery waste, or food waste from

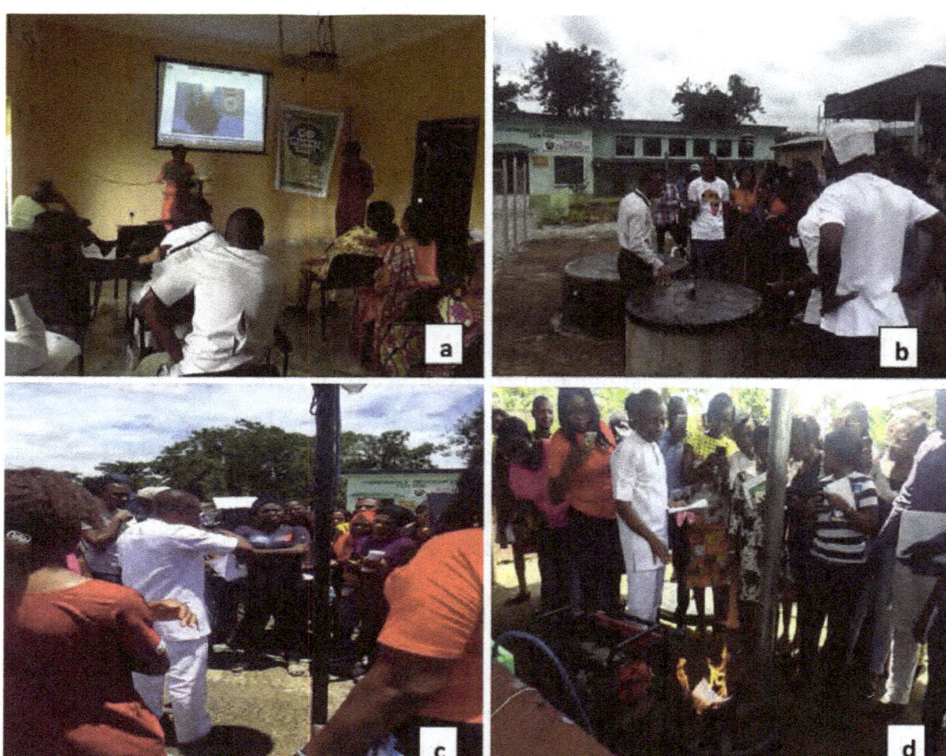

Fig. 3 Training activities at the center (**a**, **b**, **c**, **d**)

Fig. 4 Various components at RRC: (**a**) biogas compression unit; (**b**) effluent treatment filtration tank; (**c**) mushroom spun production room and (**d**) con-planted on treated effluent with many fruits on a stem

kitchens. Available end products from this section are biogas lamp, biogas compressor, biogas- multipurpose electricity generator, space heater, cooking stoves, and liquid and solid fertilizers produced from an effluent treatment filtration tank (Fig. 4b). The lamp is suitable for lighting an area without greenhouse gas emissions while the biogas cleaning unit is used to increase the percentage of methane. There is a kit that can be fixed to a household petrol-driven generator so that it can be adjusted to use petrol, natural gas, or biogas. The biosolids obtained after proper filtration are made into a very rich organic fertilizer that is used for mushroom cultivation (Fig. 4c) and raising crops at the demonstration plot (Fig. 4d). Besides, the space heater is used for warming young chickens in the poultry pens. At the same time, biogas fuel conversion kits and the portable cylinder are being tested to run vehicles, including cars, motorcycles, and tricycle, on biogas production from organic waste.

Mushroom Cultivation

A mushroom hut was built using materials ideal to maintain the required low temperature and humidity. In this hut, mushrooms will be grown from waste materials such as spent slurry, sawdust, etc. Also, bottles of spun (mushroom seedlings) of various tasty varieties are produced for household cultivation.

Wood Charcoal Kiln

A lot of trees fall due to heavy breeze and storms, and they are moved out by needy people for cooking purposes in an environmentally unfriendly manner that exacerbates climate change effects. A modern kiln available at the center can convert wood and cutoff trees to quality charcoal good commercial value (Fig. 5f). The furnace

Fig. 5 Other climate change adaptive technologies at RRC: (**a**) RDF plant; (**b**) smoke-free incinerator; (**c**) household composter; (**d**) smokeless charcoal kiln; (**e**) smokeless charcoal pellets; (**f**) wood charcoal

eliminates the traditional rural practice of digging holes all around and burning the wood openly, thus releasing greenhouse gases. In a 48-h cycle, 15–25 bags of charcoal are produced. Apart from a brick-made kiln for wood charcoal, there is another metal type (Fig. 5d) that converts light putrescible wastes to smokeless charcoal that is very popular in most of the western countries. By definition, when the charcoal is burned, no smoke is generated in the kitchen. This technology converts agroforestry residues: grass clippings, fallen leaves, maize stalks, lawn mowing, paper, sawdust, and other wastes from farms to biochar useful for farm input, wastewater treatment domestic cooking (Chang et al. 2011), warming house during cold weather, septic tank treatment, and others. The end product is in the form of charcoal powder (biochar) which is pelletized using a hand-operated extruder after some binder is added. Rejects and dust in the wood charcoal kiln are charged with treated effluent from the biogas digesters to produce nutrient-rich biochar that is sold to farmers as cheap fertilizer. The center also provides biocide-embedded smokeless charcoal for malarial control and energy for domestic cooking, at the same time (Fig. 5e). In the charcoal production process, a lot of heat is lost to the environment. The technology incorporated piped water into the furnace, whereby the temperature can be used to produce hot water. The poultry on the farm uses a lot of hot water, and it is a benefit.

According to Ana et al. (2013), global warming and its consequences on climate are among the problems posed by overdependence on fossil oil reserves. These drawbacks require exploring alternative energy sources such as biofuels that are environmentally friendly and renewable. Meanwhile, Bouros and Samiou (2001), Da Costa et al. (2004), and (Karve 2006) have earlier linked gaseous emissions from wood charcoal fires especially in an indoor environment to the incidence of many respiratory illnesses. The use of smokeless charcoal can reduce indoor air pollution, diseases, and gross environmental problems, arising from solid waste mismanagement while creating access to business opportunities for women and disadvantaged groups.

High-Temperature Incinerator

For managing healthcare wastes, a recommended method is incineration operated at 1200 °C (Fig. 5b). However, none of the incinerators utilized in developing countries meets that requirement as they generally produce temperatures below 600 °C. World Bank tried to promote an incubator designed in the UK (De Montfort). Even this model is not able to maintain such an extreme temperature. Through locally developed and improved technology, the center designed an incinerator that uses heat generated from syngas. The heat is injected back into the incineration chamber to increase its temperature to around 800–1000 °C. The unit can also be used to handle pyrolysis of scrap tires and some types of e-waste which is an emerging global environmental health importance, as these wastes have become one of the fastest-growing waste types in the world (Sharma et al. 2012). The available end products in this section are carbon black and black oil that are generated from the smoke that might contribute to global warming as well as metal scrap salvaged from scrap tires. The distinctiveness of the plant is that: the unit is entirely covered to reduce the emission of GHGs, heat produced during the operation, in form of syngas, is used to replenish the furnace temperature, and the smoke generated is converted to diesel after some treatment. The unit can be operated safely at any convenient place since there is no pollution of the environment. Additionally, materials used in constructing the unit are locally available while the unit powers itself to treat hazardous wastes, hospital waste, and agricultural wastes.

Refuse Derived Fuel (RDF) Plant

This compartment uses small wood chips (paper or any dry waste) and charcoal which on heating produce a black oil and, on further heating, produce a gas (syngas). This gas can be stored in a bag or cylinder and may be used to run a household generator to produce electricity, similar to biogas. The gas can also be used directly such as biogas for cooking at the household or institutional level. In this section, the end products are syngas and charcoal that emit very little smoke. This new innovative technology (Fig. 5a) has been developed to generate syngas from all types of combustible solid wastes and use the gas to run generators for electricity. In addition to health safety and environmental protection, this technology will be of great benefit to the rural and peri-urban populations who may want to develop small or medium scale entrepreneurship with low investment and very little skills. It does not need electricity or any sophisticated mechanical devices.

Gasification- and pyrolysis-related processes are usually confused with one another. Zhou et al. (2008) identified the major difference between the two methods as the gasification occurs in the presence of oxygen. On the other hand, pyrolysis does not require oxygen or air during its process. However, gases and liquids produced in the pyrolysis and gasification processes have high contents of carbon monoxide (CO) and hydrogen gas (H_2) that can be a good source of syngas. The syngas is used for the production of methanol and ethanol or H_2 used in fuel cells (Baumlin et al. 2006). For example, gases with the main benefit of pyrolysis is that it offers clean heat, which is needed to develop cooking technology with lower indoor pollution by smoke than is typically generated during the burning of biomass (Bailis et al. 2005). Also, the by-product is smokeless briquette charcoal, because, during carbonization, its smoke disappears.

Home Composter

This unit was designed for low-scale production of compost at houses and farm units (Fig. 5c).

The RRC is intended to be operated on a sustainable basis through various sources of revenue generation, including the following:

(a) Consultancies and turnkey projects for public and government bodies on the technologies developed at the center (fertilizer, biogas, charcoal, smokeless charcoal, etc.)
(b) Sale of products such as charcoal, biogas kits, mushrooms, maggots, liquid and solid organic fertilizers, black oil, carbon black, and organically grown produce from the demonstration plot

(c) "Science Tourism" and education visits of schools, and the public
(d) Assisting researchers through the use of locally fabricated equipment
(e) Training programs (short-term) and hands-on training
(f) Donors and philanthropists
(g) University support for innovative researches and grants
(h) Hospital waste management services
(i) Compression of gas into cylinders for domestic uses

From Climate Change to Community Change: Lesson from Kube-Atenda, Nigeria

This scheme is community-led waste management for reducing greenhouse gas emission and mitigating climate change effects at Kube Atenda, Ibadan, Nigeria. The project was carried out under the Climate Impacts Research Capacity and Leadership Enhancement in Sub-Saharan Africa (CIRCLE) program and executed by the UK's Department for International Development (DFID), in collaboration with the Association of Commonwealth (ACU) and African Academy of Sciences (AAS). Transformative adaptation and sustainable responses to climate change (Fig. 6a–d) that were adopted included: a large scale stakeholders' workshop that comprised community people (the users), local council representatives, private practitioners, and selected people from the general public; creation of a sorting center/recycling kiosks for recyclable buyback arrangement and entrepreneurship opportunities; Waste-to-Electricity through a newly invented pyrolytic chamber for converting accumulated waste to energy and useful by-products; developing market links for recyclables between the community's buyback center and resource recovery centers; developing effective communication among all the stakeholders and decision-makers; and instituting monitoring and evaluation for progress control through the involvement of Department of Community Development Inspectorate (CDI) at Ibadan Northeast LGA (local council). Specifically, this project focused on behavioral change of the community member toward solid waste management practices that will ensure wealth creation and development, using the 4R concept (Reduce, reuse, recycle, and recover).

Before this uptake activity, the community had been faced with a big challenge of solid waste management. There were heaps of waste around the community while people were disposing of their garbage on illegal dumps, in open spaces, and in water bodies, or burning them openly in an environmentally friendly manner. The training component of the research led to building people's capacity in waste to wealth as the community members were taught on how to sort their solid waste into paper, glass, plastics, and other recyclables for recycling, reuse, and sale at the community buyback center. They were also taught a way to produce organic fertilizer from organic waste and biochar (smokeless charcoal) from agroforestry residues to generate an inexpensive and clean source of energy for domestic cooking. There was a rise within the level of awareness of the consequences of poor solid waste

Fig. 6 Climate change mitigation and adaptation initiatives at Kube-Atenda: (**a**) recycling kiosks for recyclable buyback arrangement, (**b**) field demonstration on charcoal production, (**c**) pyrolytic chamber, and (**d**) phone-charging center

management, such as health hazards, esthetic damage, and other environmental problems in the community.

Moreover, the pyrolytic chamber generates gas that powers a generator for electricity supply and produces black oil from which diesel and kerosene can be distilled. The electricity supplied is used to power a phone-charging center (Fig. 6d) that financially empowers the youths who are managing the center.

The center is now a study area for young- and midcareer researchers and postgraduate students in universities across the country. This project is meeting the people's needs, including economic development, safe environment, poverty reduction, and healthful living. Also, policymakers in the state and across the country will find the uptake and research evidence very useful in their planning and activities.

Integrating Waste Picking at Dumpsites into a Value-Added Chain in Nigeria

Old dumpsites contain a high proportion of recyclables, including minerals and metallic components including e-waste, ferrous scrap, aluminum, and copper cable that could be salvaged and reprocessed. Also, the highly calorific waste components (wood, plastics or paper, textiles, and paper) could be used to generate energy (Bachmann and Cordes 2007). Recycling of these wastes has the potential to create jobs for waste pickers, by promoting entrepreneurs to establish waste recovery facilities at dumpsites. Adding waste into the value-added chain formally will reduce the quantity of wastes disposed to landfill sites and improve the practices that endanger the health and safety of waste pickers, at landfills/dumpsites.

In many developed countries such as those within the European Union, policies of reduction, reuse, and diversion of wastes from dumpsites/landfill are strongly promoted (European Environment Agency 2003). The situation is quite different in Nigeria as the main fractions of wastes generated are recyclables including organic materials, plastics/rubbers, cardboard, metals, and glass bottles (Ojolo 2004). These wastes are stored and transported directly to some designated dumpsites where they are burnt eventually. This practice has potential impacts on the health and hygiene of people in the neighborhood and the esthetic nature of the environment.

Zerbock (2003) noted that waste dumping is the well-liked method of putting off solid waste rather than sending wastes to landfills in most African countries. Wastes are solely dumped in low-lying areas without engineered measures to prevent environmental pollution. In open dumps, there is no consideration for leachate and gas management. Also, operational standards, such as registration of users, control of the number of tipping fronts, or compaction of waste, are few, if not absent. This method is neither hygienic nor safe as waste is tipped haphazardly. However, African countries have a minimal choice because of their low financial capacity and lopsided institutional arrangements. The responsibility of waste disposal is vested in local governments which are weak and cannot usually raise money to construct sanitary landfills. Some countries use semicontrolled landfills whereby wastes are dumped uncontrollably at designated dumpsites without covering with topsoil to prevent odor and rodent nuisance. In this kind of landfill, various wastes such as municipal, industrial, or clinical/hospital wastes are dumped, not following internationally approved best practices.

Waste-Picking Activities at the Dumpsites: A Case Study of Ibadan, Nigeria

Through Focus Group Discussion (FGD) sessions and observation checklist, the authors assess the dynamic of waste picking at the four designated dumpsites in Ibadan, Nigeria. The dumpsites included in the survey were: Ajaganka, Awotan, Lapite, and Afonfura. For the FGD, questions were structured to cover the following major activities: population and category of workers, business activities, method of business apprentice, factors that affect the operation, including technical, environmental, social, and economic factors, and perception of scavengers on business optimization and improvement.

Category of Workers and Prices of Items

When asked about the number of scavengers at each dumpsite, the following responses were gathered: Ajaganka – 20 (17 males and 3 females); Awotan – 20 (15 males and 5 females); Lapite – 5 (4 males and 1 female); and Afonfura – 28 (18 females and 10 males). There were more males than females except at Afofunra where 18 females were found as against 10 males. As per the year of experience, the age ranged between 3 and 20 years. According to the discussants, there was a competitive demand for recyclables. They sold the recyclables to the manufacturers who came to buy them directly from the sites. Each recyclable had a fixed price and the unit of measurement was either in kg, ton, or dozen, depending on the types of recyclables. The prices of the recyclables were given thus: tin (including milk can) – 20/kg; pet bottles – 30/kg; glass bottles – 20/dozens (depends on sizes and color); plastic scrap – 30/kg; nylon – 12/kg; aluminum (including some beverage cans) – 50/kg; copper (wire) – 800/kg; brass – 400/kg; steel – 30/kg; zinc – 50/kg; stainless metal – 80/kg; electronics (including computer panel) – 800/kg; sacs – 60/dozen; carpet – 20/kg; and Rug – 10/kg. The prices of the recyclables depend on demand from industrialists who use recyclables as potential materials for manufacturing. Economic aspects of salvaging recyclables from dumpsites have been widely reviewed (Rettenberger 1995; Van der Zee et al. 2004). In these studies, possible costs and benefits that may be obtained from using dumpsite as material recovery centers were outlined on a large scale. Some of the benefits listed were revenue from recycled materials and saving land space at dumpsites to increase their life spans. Contrary to this observation, the economic feasibility of landfill mining could not be meticulously addressed by previous studies (Savage et al. 1998; Cossu et al. 1996), and in some studies, such initiatives have been considered nonbeneficial (Cobb and Ruckstuhl 1988; Bryden 2000).

At all the dumpsites, more males were found than females because more physical strength required for the job was attributed to males and that females were usually discouraged by the stigmatization attached to the work by the people. The discussants were satisfied with their job because of their achievements and sustainable daily income from the sale of recyclables.

Method of Business Apprenticeship

The waste pickers could not identify any method of apprenticeship during the discussion. The majority of them joined the business haphazardly; they associated their enrolment into the job with severe financial problems and bankruptcy. Their main plan was to leave the job as soon as they recovered from insolvency. Contrarily, having tasted the good returns from the job, they found it difficult to go, even after they had overcome their financial burdens. The waste pickers said further that the business required little or no capital to start with and the principal instrument required was a picker. They did not register with any government agency or ministry. As such, there was no monitoring from the government or any nongovernment organization.

Factors That Affect the Operation of Waste Pickers

Among other things, almost all of the discussants listed the following as significant constraints they faced in the job: lack of proper and controlled tipping that sometimes leads to site congestion and open burning, disturbances from environmental health officers who usually extort money from them, odor and flies from abattoir waste, transportation, lack of business regulation, and fire outbreak. Other constraints cited were sickness (mostly from fever, body pains, whitlow, and laceration) as well as smoke, and flies and odor problems. When asked about steps taken for their preventive measures, many of them used self-medication. Besides, almost all of the discussants did not use Personal Protective Equipment (PPE), claiming that the use of PPE was not convenient to them and that they lacked financial capability to do so. This practice may have a profound effect on their health and well-being. Most of the waste composition studies address environmental and safety issues at dumpsites (Hogland et al. 1995; Cossu et al. 1995; Zhao et al. 2007; Prechthai et al. 2008) and show that the practices at municipal dumpsites are not effective. Many of the dumpsites have no fence, allowing unregulated access to the site and dumping of restricted materials, such as toxic wastes. Additionally, owing to the poor control of activities at dumpsites, medical and hazardous wastes end up at those dumpsites even though there are no particular dumping areas for such wastes in the city.

Perception of the Waste Pickers on Business Optimization and Improvement

Suggestions on how the job can be improved were sought from the waste pickers. Very few of them that responded said the government should absorb them into the civil service, give them proper training, especially on waste recycling, and support them with necessary facilities. Also, the discussants thought that the marketing of the recyclables could be improved significantly if the government could purchase the recyclables from them and act as a wholesaler. According to Mocker et al. (2009), due to the increasing scarcity and costs of different raw materials, the recovery of recyclables from dumpsites as a source for raw materials is rising steadily.

Quantity of Waste Picking Per Month at the Four Dumpsites

The output, in terms of quantities of recyclables salvaged per month for the four sites, is represented in Fig. 7. The most purchased component was plastics, followed by paper and organic waste. Rugs and electronics were found at the lowest proportions. The thin film and organic waste were not sorted, though they equally had high recycling potentials. According to the participants, "*Organic waste was messy and very difficult to deal with*"; *it could not be stored without much problem, especially those related to odor and flies nuisance.* Besides, thin film such as pure water sachets remained very common; it could be found anywhere and everywhere without stress, leading to its low offered price. Waste components such as disused mattresses and packaging plastic foams had very little or no secondhand values; they could instead be reused than recycled. Also, an individual waste picker was assigned with a specific type of recyclable, and the number of waste pickers working at each site reflected the quantities of recyclables obtainable at such site.

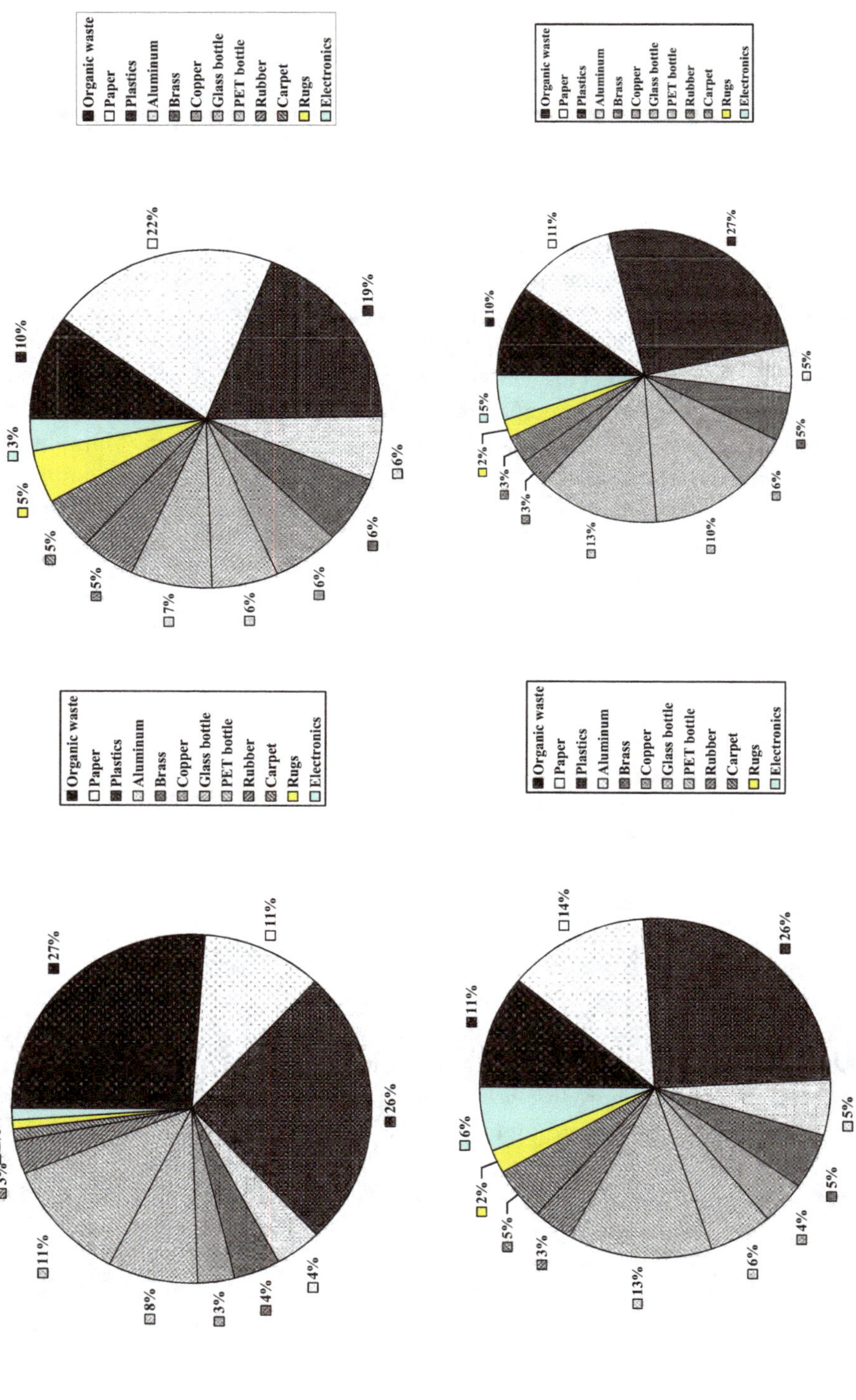

Fig. 7 Quantity of recyclables at the four dumpsites per month. Left top (Lapite); Left bottom (Afonfura); Right top (Awotan); Right bottom (Ajakanga)

Conclusion

The rate at which climate change affects developing economies has called for urgent measures to cope with unexpected climate phenomena. Climate change effects threaten to exacerbate existing vulnerabilities and pose unprecedented challenges and opportunities for people in this part of the world. Based on these premises, this chapter describes grassroots projects and the diversity of approaches to climate change management. Apart from the results of research and field projects, this section also showcases practice-oriented procedures, case studies, innovative local adaptation practices as well as barriers and limits to climate change management. The contents are expected to foster resilience and capacity building among communities in developing countries.

References

Ana PGP, Betania HL, Rubens MF (2013) Application of biomass to hydrogen and syngas production. Chem Eng Trans 32:589–594

Ashton W, Chertow M, Shenoy M (2009) Industrial ecology – developing systemic solutions to climate change and other environmental challenges in Indian industry. Sustain Tomorrow 4(4): 48–55

Bachmann M, Cordes M (2007) Das Rohstoffpotential von Altdeponien aus wirtschaftlicher Sicht (the raw material potential of old dumpsites). In: Wiemer K, Kern M (eds) Bio- und Sekundärrohstoffverwertung II. Witzenhausen-Institut für Abfall, Umwelt und Energie GmbH, Witzenhausen, pp 857–869

Bailis R, Ezzati M, Kammen DM (2005) Mortality and greenhouse gas impacts of biomass and petroleum energy futures in Africa. Science 308:98–103

Baumlin S, Broust F, Bazer-Bachi F et al (2006) Production of hydrogen by lignins fast pyrolysis. Int J Hydrogen Energy 31:2179–2192

Bouros D, Samiou MF (2001) Short-term effects of wood smoke exposure on the respiratory system among charcoal production workers. CODEN CHETBF 119(4):1260–1265

Bryden G (2000) Factors influencing the feasibility of wood products in landfill waste recovery for beneficial re-use. In: Proceedings of the 2000 TAAPI international environmental conference & exhibition, Denver, pp 527–535

Chang ACC, Chang HF, Lin FJ et al (2011) Biomass gasification for hydrogen production. Int J Hydrog Energy 36:14252–14260

Chertow MR (2009) The ecology of recycling. The UN Chronicle online, special climate change issue 2009

Chertow MR, Lombardi DR (2005) Quantifying economic and environmental benefits of co-located firms. Environ Sci Technol 39(17):6535–6541

Cobb CE, Ruckstuhl K (1988). Mining and reclaiming existing sanitary landfills. In: Proceedings of the national waste processing conference, Detroit, pp 145–151

Cossu R, Motzo GM, Laudadio M (1995) Preliminary study for a landfill mining project in Sardinia. In: Proceedings Sardinia 95, fifth international landfill symposium, Cagliari, pp 841–850

Cossu R, Hogland W, Salerni E (1996) Landfill mining in Europe and the USA. ISWA Tear Book, pp 107–114

Da Costa JL, Navarro A, Neves JB et al (2004) Household wood and charcoal smoke increases the risk of otitis media in childhood in Maputo. Int J Epidemiol 33(3):573–578

Dehoust et al (2005) Status report on the waste sector's contribution to climate protection and possible potentials. Commissioned by the German Federal Environmental Agency, Berlin

European Environment Agency (EEA) (2003) Europe's environment: the third assessment. Environmental assessment report no 10, vol 7. EEA, Copenhagen, pp 151–164
Fauzul RS, Viradin Y (2011) Community participation as community-based integrated waste management. International Journal of Academic Research 3(2):434–441
Gwatkin DR, Rutstein S, Johnson K et al (2000) A. Socio-economic differences in health, nutrition, and population. HNP/poverty thematic group. World Bank, Washington, DC
Hammed TB, Sridhar MKC, Olaseha IO et al (2012) Community perceptions on a government-provided integrated-waste recycling plant: experience from Ibadan, Nigeria. Int J Sci Soc 3(3): 145–158
Hammed TB, Shem OW, Yacob M et al (2018) Improving knowledge and practices of mitigating green house gas emission through waste recycling in a community, Ibadan, Nigeria. Waste Manag 81:22–32
Harris P (1999) On charcoal. Interdiscip Sci Rev 24:301–306
Hogland W, Jagodzinski K, Meijer JE (1995) Landfill mining tests in Sweden. In: Proceedings Sardinia '95, fifth international landfill symposium, Cagliari, pp 783–794
IPCC (2006) IPCC guidelines for national greenhouse gas inventories. In: Eggleston HS, Buendia L, Miwa et al (eds) Prepared by the national greenhouse gas inventories programme. IGES, Japan
Karve AD (2006) Briquette charcoal from sugarcane trash. Drupal, Belgium
McDonald S, Ball R (1998) Public participation in plastics recycling schemes. Resour Conserv Recycl 22:123–141
Mocker M, Fricke K, Löh I (2009) Urban Mining – Rohstoffe der Zukunft (Urban mining – raw materials for the future). Müll und Abfall – Fachzeitschrift für Abfall- und Ressourcenwirtschaft 42(10):492–501
Ojolo SJ (2004) Conversion of municipal solid wastes into medium grade fuel and industrial raw materials in Lagos Island. PhD thesis, University of Ibadan
Oluwole O, Otaniyi OO, Olopade CO (2012) Indoor air pollution from biomass fuels: a major health hazard in developing countries. J Public Health 20. https://doi.org/10.1007/s10389-012-0511-1
Paula U, Elizabeth R (2008) Factors influencing households' participation in recycling. Waste Manag Res 26(2):140–146
Prechthai T, Padmasri M, Visvanathan C (2008) Quality assessment of mined MSW from an open dumpsite for recycling potential. Resour Conserv Recycl 53:70–78
Rettenberger G (1995) Results from a landfill mining demonstration project. In: Proceedings Sardinia '95, fifth international landfill symposium, Cagliari, Italy, pp 827–840
Salvi S, Barnes P (2010) Is exposure to biomass smoke the biggest risk factor for COPD globally? Chest 138(1):3–5
Savage GM, Diaz LF, Golueke GC (1998) Guidance in landfilling waste in economically developing countries. US Environmental Protection Agency, Cincinnati, p 467
Sharma P, Fulekar MH, Pathak B (2012) E-waste- a challenge for tomorrow. Res J of Rec Sci (3):86–93
Smith AK, Brown S, Ogilvie K et al (2001) Waste management options and climate change. Final Report ED21158R4.1 to the European Commission, DG Environment, AEA Technology, Oxfordshire, 205 pp
Sridhar MKC, Hammed TB, Anthony O (eds) (2016) Entrepreneurship in waste recycling. Book Builders, Ibadan. 750 pp
The Guardian (2015) UNHabitat sets seven-point agenda to transform Nigerian cities. https://guardian.ng/property/unhabitat-sets-seven-point-agenda-to-transform-nigerian-cities/. Accessed 25 Feb 2020
Tucker P, Lamont J, Murray G (1998) Predicting recycling scheme performance: a process simulation approach. J Environ Manag 53:31–48
UNEP (2010) Waste and climate change: global trends and strategy framework. United Nations Environmental Programme Division of Technology, Industry and Economics, International Environmental Technology Centre, Osaka/Shiga

US Environmental Protection Agency (2006) Solid waste management and greenhouse gases: A life cycle assessment of emissions and sinks, 3rd ed. US Environmental Protection Agency, Washington

Van der Zee DJ, Achterkamp MC, de Visser BJ (2004) Assessing the market opportunities of landfill mining. Waste Manag 24:795–804

Zerbock O (2003) Urban solid waste management: waste reduction in developing nations (written for the requirements of CE 5993 field engineering in the developing world). Michigan Technological University, Houghton, pp 89–105

Zhao Y, Song L, Huang R (2007) Recycling of aged refuse from a closed landfill. Waste Manage Res 25:130–138

Zhou CH, Beltramini JN, Fab YX, LU GQ (2008) Chemoselective catalytic conversion of glycerol as a bio-renewable source to valuable commodity chemicals. Chem Soc Rev 37:527–549

10

Intersectional Perspective of Strengthening Climate Change Adaptation of Agrarian Women in Cameroon

Faith Ngum and Johan Bastiaensen

Contents

F. Ngum (✉)
Lilongwe, Malawi

J. Bastiaensen
Institute of Development Policy (IOB), University of Antwerp, Antwerp, Belgium
e-mail: johan.bastiaensen@uantwerpen.be

Abstract

It is a widely accepted notion that climate change affects men and women within agrarian populations differently; consequently, their adaptation strategies are gendered. Besides climate change, women's vulnerability and their corresponding adaptation strategies are embedded within a complex web of social identities/status, agroecological location, gender norm/roles and power struggles within the plurality of normative orders governing land (property rights). This chapter focuses on Cameroon and seeks to analyze how the interactions between various normative orders governing access to land, co-dependent upon the multiple gendered identities (intersectionality), impact climate change adaptation strategies of female farmers. The results show that the degree of vulnerability and adaptation strategies of women are context specific and gendered across the five distinct agroecological zones of Cameroon. Furthermore, secured access to and ownership over land is crucial in determining the adaptation choices and options available to female farmers. A complex mix of state and non-state norms govern property rights in Cameroon, within which women have to constantly negotiate their land claims. These negotiations are influenced by marital status, ethnicity, educational level, and community/social relations, such that the outcome translates differently for women within the Muslim, Anglophone, and Francophone communities. The chapter concludes with context-specific recommendations to strengthen the adaptive capacity of agrarian women across Cameroon.

Keywords

Women · Property rights · Intersectionality · Climate change · Adaptation · Cameroon

Introduction

Climate change and its adverse effects on the livelihoods of agrarian communities in the global south have gained visibility within global development debates. Although these adverse effects are diverse and context specific, they are projected to hit developing countries in the global south the hardest due to technological, resource, and institutional constraints to adaptation, as well as their over dependence on natural resources as a livelihood base (Ngondjeb 2013; Pumenta 2014). Furthermore, issues of gender and climate change adaptation are increasingly being raised as women are looked upon as a marginalized vulnerable group relying on agriculture and natural resource related activities for sustenance (Denton 2002; Arora-Jonsson 2011; Kaijser and Kronsell 2013). Gender roles/norms, unequal power relations, differential access and entitlements to land entails differences in the level of vulnerability, and effects of climate change on agrarian men and women, as well as their adaptive capacity. Thus, it is a widely accepted notion that climate change affects men and women within agrarian populations differently;

consequently, their adaptation strategies are gendered (Denton 2002; Arora-Jonsson 2011). However, most studies focus on the binary construction of gender (men vs. women) thereby overlooking the influence of other factors such as geographical location, social relations, power, social/societal norms, social identities, etc. on the adaptive strategies of farmers, especially women (Kaijser and Kronsell 2013; Osborne 2013; Carr and Thompson 2014; Kings 2017). Recent intersectionality approaches to gender analysis have, however, demonstrated that the gender binary needs to be interconnected with other markers of identity and social position (Kaijser and Kronsell 2013; Osborne 2013; Van Aelst and Holvoet 2016; Boltana 2017; Elum et al. 2017).

In an agrarian economy like Cameroon with five distinct agro-ecological zones where subsistence agriculture is the mainstay for about 80% of the country's population (Molua 2011; Pumenta 2014), the effects of climate change are multiple and diverse, with negative consequences dominating. With most smallholder farmers highly dependent on rain-fed agriculture, climate variability and changes are detrimental to their livelihood base and as such the agricultural sector is among the most vulnerable to the risks and impacts of climate change (Tingem and Rivington 2009; Pumenta 2014). Although agriculture is the livelihood base for most Cameroonians, women are the backbone of smallholder agriculture and food crop production, bearing most of the responsibility of household sustenance. Women constitute 51% of the entire Cameroonian population, 51.3% of whom rely exclusively on livelihood activities that are directly linked to the natural environment (Molua 2011; UICN-PC 2015; Beckline et al. 2017). Women are thus highly vulnerable to the impacts of climate change. Besides climate change, the agrarian livelihood base of these women is further complicated by a dual land tenure system within which women (based on their social identity/status and related gender norms) are constantly negotiating land access and rights, trying to mobilize the appropriate gender and customary norms that can be used to enforce their land claims (Faith 2019). Thus, women's vulnerability to climate change and their corresponding adaptation strategies are embedded within a complex web of social identities, gender roles, and power struggles over land within the various agroecological zones. Hence, the need to look at an intersectional perspective of factors that influence climate change adaptation strategies of agrarian women in Cameroon.

This chapter therefore seeks to analyze how the interactions between various normative orders governing access to land, co-dependent upon the multiple gendered identities (intersectionality), impact climate change adaptation strategies of female farmers in Cameroon. It reviews how climate change affects agrarian populations and how they generally adapt, both at the household and community level. Specific attention will then be given to women's access to farmland within the plurality of normative orders governing land and how land rights, social identities, and socioeconomic factors jointly influence their adaptation strategies (intersectionality).

This chapter is an exploratory desk study based on general literature review and discourse analysis of peer-reviewed journal articles, government policy documents, reports, and other unpublished texts about gender/women, climate change vulnerability and adaption strategies, intersectionality, and land governance. A keyword search was used to identify and isolate relevant texts, using gender/women, climate change adaptation, climate change vulnerability, intersectionality, smallholder farmers, and land tenure as keywords. Approximately 50 documents were obtained from the keyword search. In order to understand the current/recent dynamics relating women and intersectionality to land tenure, climate change vulnerability and adaptation in Cameroon, the focus was on literature (journal articles, policy documents, reports, etc.) produced within the last 10 years (2009–2019), except one document on land tenure issues in Cameroon which dates back to 1996. This document was retained because no other recent/current document was found. Thus, based on their relevance to the study, a quick scan through the 50 documents obtained led to the elimination of 15 documents. In summary, this chapter exploits literature from 35 documents on the following thematic areas:

- Gender, climate change vulnerability and adaptation strategies
- Land rights, smallholder agriculture, intersectionality, and climate change adaptation

Conceptual Framework: Intersectionality

This study adopts the concept of intersectionality that emerged as a critique of the universalism of women's oppression portrayed by initial Western feminist movements. The concept of intersectionality was first used by black feminist (specifically by Kimberlé Crenshaw) to elucidate the triple oppression and discrimination of women of color resulting from an interplay of race, gender, and class. Although the concept emanates from black feminist movements, it has evolved over the years into multiple disciplines of feminist research to illuminate the interconnectedness of power and multiple axes of irreducible social categories (beyond race, gender, and class), and their effect on the identity and lived experiences of women (Davis 2008; Cho et al. 2013; Kaijser and Kronsell 2013; Kings 2017). However, even within feminist research, intersectionality has been defined, interpreted, and used in various ways (an analytical tool, a research paradigm, a buzzword, etc.). Given the ambiguity and open-endedness of the concept, we define intersectionality as "*the interaction between gender, race and other categories of difference in individual lives, social practices, institutional arrangements, and cultural ideologies and the outcomes of these interactions in terms of power*" (Davis 2008, p. 68). This definition highlights the fact that intersectionality is inextricably linked to power analysis between an infinite combination of overlapping social identities and categories that culminates to inequalities, discrimination, and oppression within a broader societal context. Put simply, intersectionality acknowledges that oppression is the outcome of interactions of multiple simultaneously linked identities, power relations, and lived experiences. Thus, the concept of intersectionality has been applied in myriad ways within critical feminist studies to highlight the historical "situatedness," social relations, and power structures that results into the irreducible, layered and co-constitutive identities of women within in a given context (Cho et al. 2013; Osborne 2013; Hankivsky 2014).

Within the context of climate change, "*[t]he responsibility, vulnerability, and decision-making power of individuals and groups in relation to climate change can be attributed to social structures based on characteristics such as gender, socio-economic status, ethnicity, nationality, [. . .] and place*" (Kaijser and Kronsell 2013, p. 420). Consequently, the impacts and corresponding adaptations strategies of individuals and groups may reinforce or alternatively change these social structures and categories. Thus, the degree of climate change vulnerability, impacts, and adaptation strategies of smallholder farmers (especially women) is influenced by multiple intersecting social identities embedded in power structures (Arora-Jonsson 2011; Kaijser and Kronsell 2013; Osborne 2013; Carr and Thompson 2014; Hankivsky 2014; Kings 2017). Hence adopting the lens of intersectionality moves beyond "women as a homogenous analytical category" to unravel the dynamics of the interconnectedness of power, social relations, socioeconomic status and their various co-constitutive and mutually enforcing identities of women in relation to climate change vulnerability, impacts, and adaptations. This permits an informed and constructive understanding of gender and the multiple normative orders that determine the lived experiences and "true" effects of climate change on women within agrarian communities in Cameroon. An intersectional approach will therefore incorporate the concepts of gender, land tenure, and socio-economic status so as to gain critical and constructive insights on the vulnerability, impacts, and adaptation strategies of over 50% of women in Cameroon reliant on agriculture and other natural resource base activities for sustenance in different agroecological zones.

Climate Change Vulnerability and Effects Across the Different Agro-ecological Zones

In an agrarian economy like Cameroon, where a majority of the population rely on rain-fed agriculture and natural resource exploitation (especially forestry and non-timber forest products–NTFPs) for sustenance, climate change impinges on the climate-sensitive livelihood activities of most Cameroonians. Cameroon currently faces increased annual average temperatures and seasonal rainfall fluctuations with accompanying hazards such droughts, floods, heat waves, mudslides, etc. which are experienced differently across the country's agro-ecological zones, with diverse effects and impacts on the farmers. Located between West and Central Africa, Cameroon has five distinct agro-ecological zones with distinct topographic and climatic diversity which can be assumed to account for the variations in the degree of vulnerability and corresponding effects on agricultural activities within each zone (Yengoh et al. 2010; Pumenta 2014; MINEPDED 2015). Figure 1 presents five agroecological zones and their distribution across the national territory.

The Sudano-Sahelian Zone

Covering the North and Far North Regions, it is the driest part of the country with a semiarid climate. Annual temperatures here range between 17 °C and 34 °C over

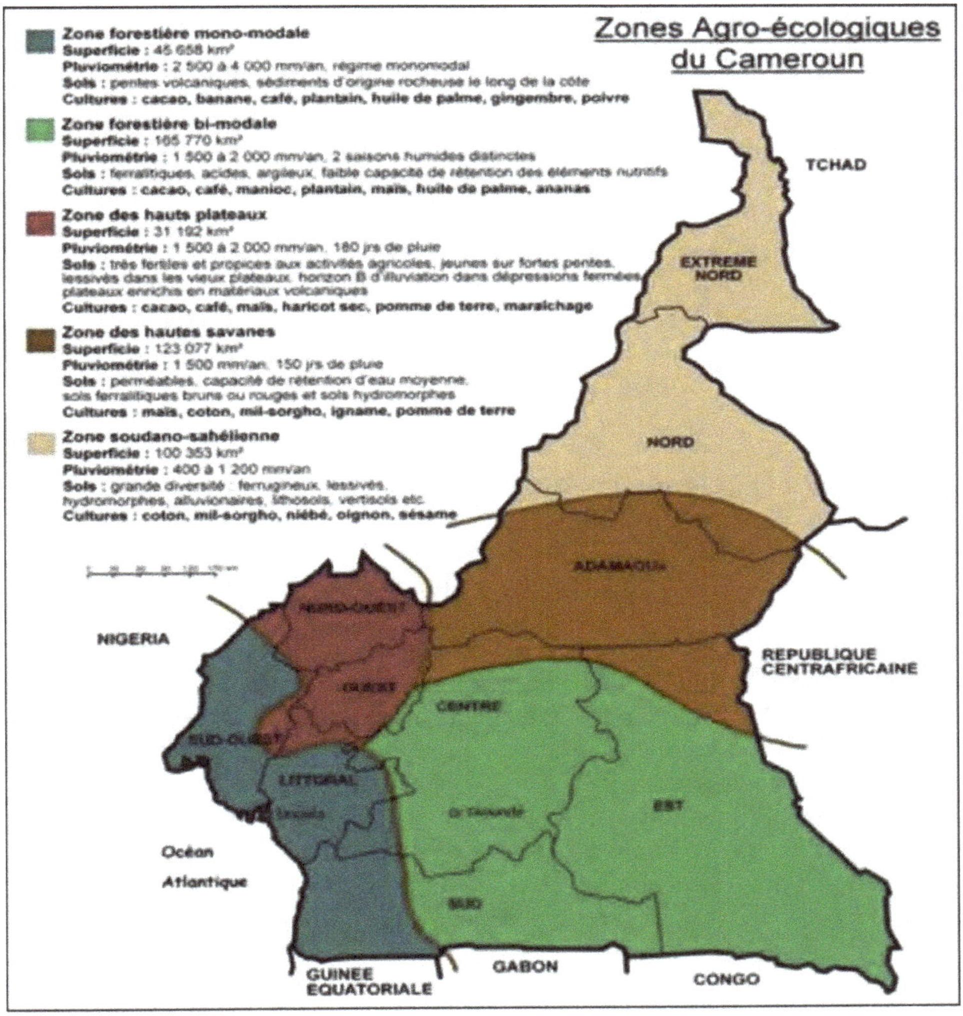

Fig. 1 Agro-ecological zones of Cameroon. (Source: MINEPDED 2015, p. 24)

long periods of the year making it highly vulnerable to droughts, heat waves, and dry spells. Despite its low rainfall distribution (between 400 mm and 1200 mm), extreme interannual variability often with strong torrential showers are common, making the regions periodically vulnerable to floods (with high incidence of waterborne diseases). It is the most vulnerable zone in Cameroon with serious repercussions on the agricultural sector. Some of the consequences of climatic variations include: drop in crop yields (resulting from poor germination, crop drying), decreased fodder coverage, increased proliferation of animal pest (especially tsetse fly) and livestock mortality, drying up of water bodies (lakes and ponds), drop in fish production, and exhaustion of fish stocks. The abovementioned consequences translate to famine and food insecurity, increased conflicts over arable land, and population migration to other regions (Yengoh et al. 2010; Pumenta 2014; MINEPDED 2015).

The Guinean Savannah Zone

It covers the Adamawa region and is a transition zone between the moist equatorial southern regions and the drier Sahelian northern regions. It is characterized by average temperatures of 22 °C and annual rainfall ranging between 700 mm and 1500 mm making it suitable for production of a variety of crops and livestock. This zone is considered the "breadbasket" of the country but with increasing temperatures, decreasing precipitation together with erratic rainfall patterns, the region is vulnerable to droughts, dry spells, and heat waves. The effects and impacts of climate variations in this zone are similar to those of the Sudano-Sahelian zone presented in the previous section, although more moderate (Yengoh et al. 2010; Pumenta 2014; MINEPDED 2015).

The Western Highlands Zone

It covers the West and North-West Regions of the country. Although it has a savannah vegetation, its mountainous topography with high plateaus (over 800 m tall) influences its climate. Influenced by this orographic effect, its annual rainfall ranges between 1500 mm and 2000 mm with an average temperature of 20 °C giving it a cold equatorial climate with the possibility of off-season rains. This region experiences the highest seasonal fluctuations of rainfall and temperature patterns. Climate variations here results in drop in crop yields (crop decay during flooding, soil leaching, and erosion), increase proliferation of animal pest (especially tsetse fly), and livestock mortality. These translate to increased competition over arable land and farmer-grazer conflicts. However, erratic rainfall patterns provide the opportunity of off-season farming and thus stabilizing crop production in this zone (Molua 2009; Yengoh et al. 2010; MINEPDED 2015; Innocent et al. 2016).

Coastal Lowlands or Mono-modal Forest Zone

It covers the Littoral and South-West mountainous regions characterized by a rainy, hot, and humid equatorial climate. It registers the highest rainfall in the country ranging between 2500 mm and 4000 mm throughout the year with mean temperatures of 28 °C. The population practice artisanal fishing and fish farming, and its rich volcanic soils, moisture-laden tropical maritime climate, and sea access makes it suitable for roots, tubers, and plantation agriculture. However, with increasing temperatures and intense rainfall, this zone is vulnerable to sea level rise, storm surges, floods, and mass movements such as landslides, mudslides, and falling rocks with negative effects on the population. Variations in climatic parameters results in increased proliferation of pest (malaria parasites), saline water intrusion and destruction of salt-intolerant crops, destruction of mangrove habitats, and increased waterborne diseases. However, increased sea temperatures also lead to improved fish habitat and changes in fish species diversity, hence a beneficial effect of climate change (Yengoh et al. 2010; Bele et al. 2013; MINEPDED 2015; Beckline et al. 2017).

The Bimodal Forest Zone

Covering the Center, South, and East Regions of the country, this zone harbors the country's dense evergreen tropical forest vegetation, with an average temperature of 23 °C and annual rainfall between 1,500 and 3,000 m. Exploitation of forest resources (including NTFPs and hunting), agriculture (mostly cocoa production and poultry farming), and artisanal fishing remain the principal activities of most households. Increased temperatures and decreased rainfall results in nonavailability of NTFP, increase in animal species mortality, and decrease in crop yields. However, occasional drying up of swamps and marshlands provides an opportunity for vegetable cultivation, which is a beneficial effect of climate change (Sonwa et al. 2012; Bele et al. 2013; MINEPDED 2015).

Farmers Household and Community Adaptation Strategies

Cameroon has a National Adaptation Plan of Action (NAPA), which mentions gender issues in reference to the need to incorporate and prioritize women (as a homogenous group) in all sectoral adaptation initiatives at all levels in the different agro-ecological zones. Although the NAPA has clearly defined adaptation initiatives developed by state institutions at national level, its effective implementation at regional and local levels is often limited by structural (poor-coordination within and across ministries), financial, and technological (outdated research and climatological stations) constraints. Thus, farmers and rural households do not rely solely on state assistance for adaptation options. Over the years, they have developed various individual and collective strategies (both on-farm and off-farm) to cope with, and lessen the adverse effects of climate change based on their local knowledge, perception, experience, and interpretation of natural phenomena. Although there is a variability among the adaptation practices of male and female farmers, in general, smallholder farmers (both male and female) adopt a combination of on-farm agricultural practices among the following (Molua 2009, 2011; Tingem and Rivington 2009; Innocent et al. 2016; Beckline et al. 2017):

- Use of improved seed varieties (drought-tolerant and early maturing varieties)
- Seasonal alterations of farm size and/or location
- Adjusting sowing/planting time (early planting in mono-modal and bimodal forest zones, late planting in Sudano-Sahelian and Guinean-Savannah zones) and sequencing (plant spacing, row orientation, terracing)
- Mixed farming, crop rotation, and agroforestry
- Manual irrigation of crops using boreholes, watering cans, and hand buckets
- Installation of sprinkler irrigation systems

- Adjusting level and timing of fertilizer and pesticide application
- Gardening on dried marshlands and swampy areas (in the bimodal forest zones)
- Implementing soil–water conservation techniques (minimum tillage, mulching, use of compost/manure)
- Decreasing herd size, increasing livestock diversity, and switching to heat- and drought-tolerant animal species
- Seasonal animal migration to river banks (transhumance)
- Increase collection and marketing of NTFPs for food and income, respectively

In addition to these agricultural practices, farmers are increasingly engaged in various off-farm income-generating activities to cushion farm income shocks such as petty trading and migration (to seek employment). Distress sale of household assets and reduction in nonfood expenditures are also observed as more detrimental coping mechanisms. Although these individual adaptive measures are successful, there is a limit to its overall contribution in a broader community resilience. Collective action is therefore needed to build community resilience against climatic shocks. Hence, the aforementioned individual adaptation strategies are complemented by other community actions which include: community rotational saving schemes ("*tontines*" or "*njangis*") with lending possibility, group collection and marketing of NTFP, cultivation of fodder crops for livestock, creation of community fodder banks, creation of community firebreaks (protecting against wild fires), family/community information sharing meetings, and community regulation of common-pool resources (land, water, and forest) (Molua 2011; Bele et al. 2013; Innocent et al. 2016).

The array of on-farm and off-farm adaptation strategies used by farmers in Cameroon vary across the agro-ecological zones and is influenced by an intersection of other factors such as gender, property rights, educational level, income (including credit access), access to technical assistance and extension services (including climate-related research and information) (Molua 2009; Innocent et al. 2016). This highlights the importance of intersectionality in climate change vulnerability and adaptation as will be discussed later.

Women and Land Tenure in Cameroon

Land governance/tenure in Cameroon is characterized by a plurality of normative orders, wherein the effective "rules-in-use" and daily management practices are determined by the interaction between colonial and post-independence state "law" (state normative order) on one hand, and the existing social/cultural norms of the people on the other hand (Belaunde et al. 2010; Pumenta 2014). A series of gender-neutral laws, decrees, ordinances, and their corresponding implementing institutions define the state normative order. Shaped by its colonial history, the state normative order, which is applicable to the entire national territory, is based on the notion of "private ownership" and land registration documented by formal instruments (title deeds) as a basis for securing property rights and economic development. Resource access and subsequent rights are therefore determined by the ability to pay (Van den Berg 1996; Fonjong 2012; Fonjong et al. 2013; Pumenta 2014; Njoh et al. 2017). This provides equal opportunity for all irrespective of gender, ethnicity, or religious background to have unchallenged rights over land. However, various financial and socio-cultural factors shape the gender-differentiated access to land.

The religious and social/cultural norms (non-state normative order) governing land are based on communal ownership, with slight differences across the country especially concerning women's land claims, as these engulf gender norms and roles embedded within unequal and changing power relations (Faith 2019). Cameroon is made of over 250 different ethnic groups spread across the ten regions of the country with diverse gender ideologies, norms and roles in relation to land, natural resource management, and agricultural production (Njoh et al. 2017; Faith 2019). Thus, as Van den Berg (1996) opines, "*ideas about breadwinnership, the division of labour and responsibilities within the household largely influence women's access to land*" (p. 18). However, these 250 ethnic groups can be regrouped under three main communities based on their religious (Christian or Muslim), linguistic (Arabic, French, or English), and colonial history, namely (Van den Berg 1996; Ahidjo 2012; Njoh et al. 2017; Faith 2019):

- Muslim community: made of the French- and Arabic-speaking Adamawa, North and Far North Regions of Cameroon, colonized by the Arabs, Germans, and French.
- Anglophone community: predominantly Christians and comprised of the English-speaking North-West and South-West Regions of Cameroon, colonized by the Germans and British.
- Francophone communities: predominantly Christians and comprised of the French-speaking South, East, West, Littoral, and Center Regions of Cameroon, colonized by the Germans and French.

Within these communities, a woman's access to and rights over land is determined by a mix of colonial, state, and non-state normative orders, influenced by other social determinants such as gender norms, ethnicity, and socio-economic status. The following sections highlights daily struggles female farmers encounter to access arable land across the different communities.

Muslim Communities

Within the Muslim communities, an important determinant of a female farmers' intra-household access to land is her marital status, as wives primarily rely on their husbands to gain access to farmland. However, access to farmland is not automatic upon marriage but earned through years of marital compliance, successful reproduction, and labor provision on the family field/farm. Wives are granted permissive use rights on degraded marginal or fallow lands for food crop cultivation, while their husbands still hold control and transfer rights over the land. With divorce or widowhood, divorcees lose access and rights over family land while widows with minor children may be granted usufruct and "limited" control rights over land held in trust of their son (the heir to the land). Nevertheless, wives, divorcees, and widows have devised alternative means to temporarily obtain farmland in their community within the appropriate socio-cultural norms used to enforce their land claims, such as renting and sharecropping. Inheritance and purchasing land could be a viable option of obtaining secured property rights but land ownership by "women" is stereotyped as a threat to family unity (decrease of sibling/son's inheritance or signs of marital discord), thus land purchase by women is almost inexistent within the Muslim communities. Furthermore, daughters and widows can only inherit land in the complete absence of a male relation (which is never the case) (Van den Berg 1996; Ahidjo 2012; Faith 2019). The insecure and temporary nature of these land tenure arrangements often limits the amount of investment Muslim women can and are willing to make on the farmland. With the advent of climate change, their adaptation strategies are therefore constrained.

Anglophone Communities

Within the household, Anglophone women also acquire land through male relations (fathers, husbands, and sons). However, differences exist in the rights and degree of control Anglophone women hold over land. Daughters are granted "permanent" use rights on family land and could inherit the land provided she remains unmarried even though sons are prioritized in relation to land inheritance. Upon marriage, wives are automatically granted usufruct rights on family land by her husband or his kinsmen, to ensure their principal duty of household food production (unlike Muslim wives who earn their access to farmland). At the discretion of the kinsmen, widows may keep their farmland as trustees for their sons (patrilineal inheritance) or nephews (matrilineal inheritance) with a greater degree for control and management rights. Divorced women are dispossessed of any rights they held over the husband's family land and return to their natal family for land or seek alternative strategies of acquiring farmland. However, although men in this community "own" the land, women hold the decision-making power and controls what is cultivated on the land, how her labor is allocated, and how the farm proceeds are used (Van den Berg 1996; Pumenta 2017; Faith 2019).

Besides relying almost exclusively on male relations for farmland, in the community sphere, Anglophone women (daughters, wives, mothers, widows, or divorcees) have also been known to "clear and claim" land as their own, as well as acquire land through renting, collective access of communal land, purchase, and registration. Irrespective of any socio-cultural and administrative bottlenecks (gender-biased norms, high cost, bureaucracy, etc.) associated with women's land purchase, Anglophone women have been able to circumvent these obstacles by using male relatives to buy land or "courting" both traditional and state authorities to buy land in their name (Van den Berg 1996; Pemunta 2017; Faith 2019). Although possible variability in terms of income levels, class, educational level, and social relations influence their land claims and rights enforcement, women (state bureaucrats, elites, and "business" women) own about 3% of the registered land in the Anglophone communities (Pemunta 2017). Furthermore, regrouping themselves into farmers organizations to access communal land and also pool resources to purchase land for collective food production and market gardening is an important strategy for which Anglophone women have proven to exert more power in bargaining and enforcing their land claims within both state and non-state normative orders (Fonjong et al. 2013; Pumenta 2017; Faith 2019). Although, variations exist in terms of access to land and security of tenure, most Anglophone women have more or less "secured" property rights and can therefore make pertinent management decisions to undertake long-term sustainable investments on their farmlands and reap the benefits. They are therefore more likely to adopt sustainable climate smart practices to cope with the adverse effects of climate change.

Francophone Communities

A Francophone woman's primary duty within the household (in accordance with socio-cultural norms) is to provide labor support on the family farm. Hence, access to land is conditioned on her "relationship" with the men in the family. Daughters assist in farming activities as well as the collection and processing of NTFPs. Upon marriage, wives acquire a share in the land of her mother-in-law or husband, on which they can cultivate food crops and vegetables (Fonjong 2012; UICN-PC 2015). Although men hold superior power over women (daughters and wives), household decisions over crop production and labor allocation are jointly made by husband and wife. Furthermore, the sociocultural norms generally protect the land claims of all farmers irrespective of gender, social identity, or ethnic origin (women can and do inherit land), although in some rare instances, widowhood and divorce may lead to loss of farmland by women (Gyau et al. 2014; UICN-PC 2015). Besides relying on family relations for land, women (wives and widows) also engage in temporary arrangements such as renting and sharecropping as alternative means of accessing additional farmland with limited management rights over these lands. Nevertheless, some financially viable women occasionally purchase land for construction or other non-agricultural developmental purposes in urban areas as evidenced in the Center Region where women (mostly state bureaucrats) own about 15% of all registered land (Gyau et al. 2014; Pumenta 2017). Thus, Francophone women generally have secured property rights that provides them with the power and means to equitably negotiate the diversification of their livelihoods and adapt to climatic changes.

In summary, despite the diverse socio-cultural norms across the country, the land rights of most Cameroonian women are co-dependent on male relations. However, the effective rule-in-use over land management as well as the intra-household decision-making power varies across these communities. Nevertheless, women contribute substantially in agricultural production and other livelihood activities, and are primarily responsible for household subsistence despite the gender-differentiated access and rights to resources.

Intersectionality and Climate Change Adaptation

Beyond the binary division of men versus women, a combination of other societal norms and socio-economic factors influence women's adaptive capacity. Consequently, an intersectional analysis is required to understand the underlying norms, power relations, and socio-economic factors that conditions the daily/seasonal adaptation practices, and strategies employed by women (Kaijser and Kronsell 2013; Carr and Thompson 2014). In Cameroon, with over 50% of the female population relying on agriculture and natural resources for sustenance (Molua 2011; Pumenta 2017), women constitute an important group for whom risk aversion is vital for household welfare. However, the degree of vulnerability and effects of climate change on the livelihood activities of women depends on their location (agro-ecological zone), the diverse socio-cultural norms that prescribe access to and control over land, and decision-making power (as presented in the previous sections). Consequently, the adaptation strategies of women will vary across the country based on an intersection of gender, ethnicity, and agro-ecological location, as well as other factors such as educational level, access to information, access to extension services (technical assistance), income (including access to credit), social capital, etc. Hence, based on the findings of the previous sections, this section analyses the effects of an intersection of gender, ethnicity (in terms of land governance and decision-making power), and agro-ecological location (and other factors) on women's adaptation options within the northern Muslim community, southern Francophone and Anglophone communities. Figure 2 highlights the spatial distribution of these communities across the different agro-ecological zones.

Women and Climate Change Adaptation Within the Muslim Community

The Muslim community of Cameroon is predominantly located in the northern semi-arid agro-ecological zones, comprised of the Sudano-Sahelian and Guinean Savannah agro-ecological zones. Climatic variations and extreme weather events such as droughts, heat weaves, dry spells, and occasional floods lead to decreased crop yields/productivity and increased population migration to other regions (mostly men) (Yengoh et al. 2010; Pumenta 2014; MINEPDED 2015). In line with their domestic/reproductive roles, Muslim women are generally involved in food crop production (on marginal lands acquired through male relations), processing and marketing of livestock and fishery products but lack control over how the proceeds are used (Van den Berg 1996; Ahidjo 2012; Njoh et al. 2017).

The temporary migration of Muslim men during periods of transhumance and in search of off-farm employment in other regions implies an increase in the number of de facto female-headed households with increases in the reproductive and productive activities of wives who are left to care for ("sick") family members while ensuring sufficient food production for the household. Although some women are engaged in petty trading, the increased workload of domestic activities limits their possibility of

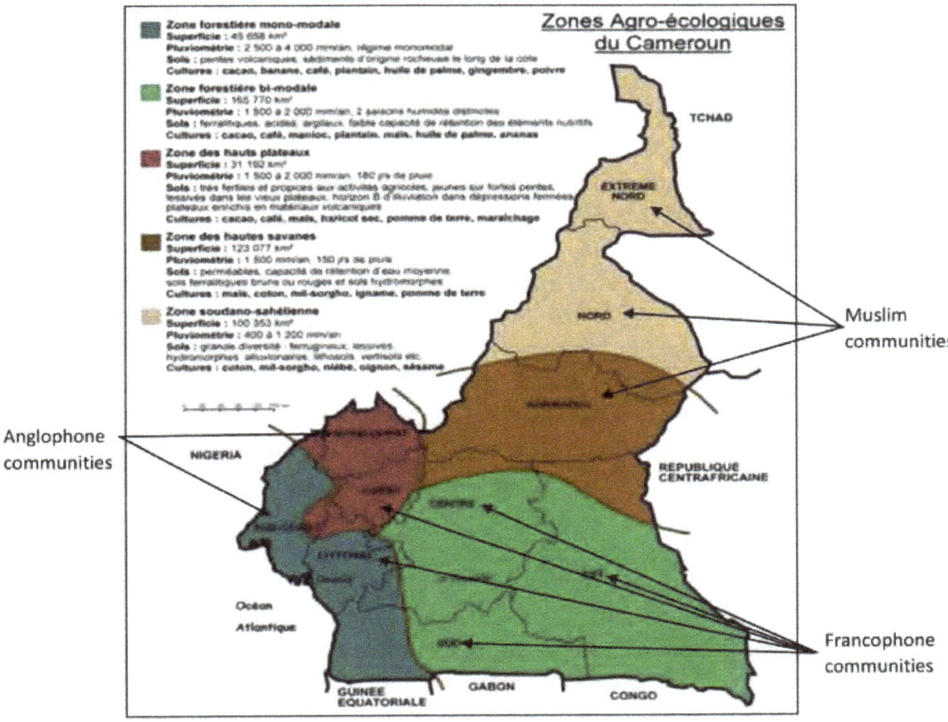

Fig. 2 Spatial distribution of the various communities across the different agro-ecological zones. (Source: Adapted from MINEPDED 2015)

diversifying their livelihood strategies and undertaking off-farm activities that could contribute to the household income. This is further exacerbated by the high illiteracy level among Muslim women, due to the gender ideology that prioritizes early marriage over girls' continuous education, thereby limiting their options of off-farm employment (Van den Berg 1996; Pumenta 2014). Furthermore, men generally hold decision-making power over the wife's labor and income allocation. Thus, for most Muslim women (wives, widows, mothers, and divorcees), their livelihood activities and adaptation strategies are centered around on-farm activities.

The on-farm adaptation strategies of Muslim women are constrained by religious and sociocultural norms that "forbid" ownership of land by women. Since women (daughters, wives, widows, and divorcees) only have usufruct rights on family or rented/sharecropping land, and "limited" decision-making power, their on-farm adaptations strategies are therefore limited to seasonal crop management practices and farming operations such as alterations in planting dates, use of drought tolerant or early maturing hybrid seeds, and mixed cropping. Furthermore, because of the limited property rights and temporal tenure arrangements (renting or sharecropping), Muslim women are unlikely to cultivate cash crops (except millet) that could increase their income (Van den Berg 1996; Molua 2011, 2012). The exception seems to be the case of widows with minor sons who have "limited" control rights over land held in trust for their sons. These widows can therefore undertake long-term investments such as tree planting (agroforestry) and soil management to safeguard the land while ensuring continuous household food production. As Van den Berg (1996) highlights, tree planting was used by widows as a means of securing and enforcing their land claims.

Beyond security of tenure and decision-making power, the choice of on-farm adaptation strategies implemented varies among women and is influenced by other factors such as income (to purchase drought tolerant or early maturing hybrid seeds), access to climate information and technical assistance. While farmers may generally have knowledge of the weather forecast and expected start of farming seasons (Molua 2011, 2012), Muslim women generally have limited access to agricultural extension services, as these services often target cash crop producers who are predominantly men. This limits the effective implementation of State's adaptation initiatives as stipulated in the country's NAPA, by sidelining the female population of farmers. Furthermore, while promoting tree planting in semi-arid zones is an essential adaptation initiative of the State (MINEPDED 2015), the limited property rights of Muslim women is a disincentive to engage in tree planting as they may not reap the benefits agroforestry provides in the long run. Thus, a gap exists between the national institutional framework and realities on the ground. Together, the afore-mentioned factors coupled with the limited livelihood diversification strategy of women and socio-cultural constraints on their adaptive capacity tend to increase the household and community vulnerability, thereby accentuating poverty in the regions.

Women and Climate Change Adaptation Within the Anglophone Community

The Anglophone community is comprised of the North-West and South-West Regions of Cameroon spread in two different agro-ecological zones. The North-West Region (NWR) is located within the western highlands zone, while the South-West Region (SWR) is located within the mono-modal forest zone. Climatic variations in the NWR leads to soil degradation (through leaching and erosion) and consequently drop in crop yields. In the SWR, extreme weather events, such as floods, lead to increased proliferation of malaria parasites, high incidence of waterborne diseases, saline water intrusion and destruction of salt-intolerant crops (MINEPDED 2015; Innocent et al. 2016; Beckline et al. 2017). In line with their reproductive roles, ensuring household food security is the responsibility of Anglophone women. Thus, farming (food crop and vegetable production) is the main livelihood activity of most women in these communities (Van den Berg 1996; Fonjong et al. 2013).

Health issues resulting from malaria parasites or waterborne diseases implies an increased workload on the domestic/reproductive roles of women (wives, mothers, widows, and divorcees) to nurture and care for the sick family members. This also increases their workloads in agriculture since they are tasked with ensuring household food security. However, Anglophone women as wives, widows, and divorcees decide and control how their labor or any proceeds from their farms is allocated (Van den Berg 1996; Pumenta 2017). They therefore have the possibility of diversifying their livelihood strategies and engaging in other flexible off-farm activities to generate income that could caution the adverse effects of climate change on their agrarian livelihood base. This is further facilitated by their relatively high literacy level (compared to the Muslim women) that enables them gain off-farm employment. From the researcher's lived experience, the high literacy level within Anglophone women may be considered the result of a trade-off wherein the education of female children serves as compensation for their limited inheritance rights.

Although an interaction of sociocultural norms and state "laws" dictate women's access to land and other natural resources, some Anglophone women have been able to manoeuver and "forum shop" from both systems to acquire more or less "secured" property rights over their farmland. While "high" income women such as state bureaucrats, elites, and "business" women have been able to purchase, title, and effectively own land, rural women with limited financial capital rely on their social networks/capital to regroup themselves and gain "secured" access to community land. Furthermore, by virtue of being the "breadbaskets" of the family, women generally have secured access to "family" land whether as a daughter, wife, mother, or widow (Fonjong et al. 2013; Pumenta 2017). Thus, the security of tenure provides an incentive to adopt and invest in sustainable soil management and agricultural practices such as agroforestry, recycling of dead organic matter, mulching, and reap the long-term benefits. Nevertheless, Anglophone women also engage in seasonal farming practices such as making "mounds," mixed cropping, off-season replanting, use of improved seed varieties, changing planting dates, etc. as coping strategies to the fluctuating climates in these zones (Beckline et al. 2017).

Beyond security of tenure and decision-making power, access to climate information and agricultural extension services also influence the on-farm adaptation strategies implemented by Anglophone women. Although both women and men get access to climate information from formal (weather observatory officers) sources through radios, print media, and televisions, as well as informal sources such as friends, family, and self-help groups (Molua 2011; Fonjong et al. 2013), access to agricultural extension services is correlated to group membership, social status, and class. While extension services target farmers groups/cooperatives (irrespective of the gender composition of the groups), "working class" women have the possibility of networking with and obtaining assistance from extension agents. Thus, being a full-time "housewife" and belonging to no farming group impinges on a woman's access to technical assistance. However, most Anglophone women belong to self-help and community saving groups ("njangis") which serves as a buffer against financial crisis in times of disasters and emergencies on one hand, and as a platform of socialization, information sharing, and peer learning within the communities on the other hand (Molua 2011; Fonjong et al. 2013; Innocent et al. 2016). Thus, Anglophone women seem less vulnerable and more proactive in adopting conservation and protective on-farm practices compared to their Muslim counterparts.

Women and Climate Change Adaptation Within the Francophone Community

The Francophone community is made of five administrative regions spread across three different agro-ecological zones. It is comprised of the West Region (located in the western highlands zone), the East, South, and Center Regions (located in the bimodal forest zone), and the Littoral Region (located in the mono-modal forest zone). Seasonal weather fluctuations in the western highlands and bimodal forest zones lead to soil erosion and drop in crop yields, and increased scarcity of NTFPs. In the mono-modal forest zone, sea level rise and extreme weather events, such as floods, leads to increased proliferation of malaria parasites, high incidence of waterborne diseases, saline water intrusion and destruction of salt-intolerant crops (Yengoh et al. 2010; Sonwa et al. 2012; Bele et al. 2013; MINEPDED 2015).

In addition to their reproductive roles, a Francophone woman's main duty as a daughter or wife is to provide labor support on the family farm. Although cash crop (mainly cocoa) production and fishing by men are the main livelihood activities, Francophone women (wives, mothers, divorcees, and widows) are involved in food crop and vegetable production on farmland acquired through diverse means. With both the state and non-state normative orders protecting the land and natural resource claims of women, the security of tenure provides women with the power to negotiate the diversification of their livelihoods. Being educated and earning an income is a bonus as it influences the intra-household bargaining power of women. Women therefore engage in diverse off-farm employment and income-generating activities such as petty trading of food crops, collection, processing and marketing of NTFPs, as well as processing and marketing of fishery products (Sonwa et al. 2012; Brown and Sonwa 2015; UICN-PC 2015; Beckline et al. 2017; Pumenta 2017). This off-farm income helps caution against the negative effects of changing climatic conditions on the agrarian livelihood base of the household. Furthermore, Francophone women engage in precautionary loan-savings schemes ("tontines") which provides women with easy and flexible access to credit that is used for on-farm and/or off-farm income-generating activities. This contributes in enhancing household income and general welfare, as well as provide financial insurance in times of disasters and emergencies. Francophone women also engage in community labor-sharing groups that provides mutual labor support on their farm. Since membership to these are based on trust, solidarity, and reciprocity, these groups are indicative of the social capital within this community, which can serve as an entry point for distribution of information, peer learning, and technology adoption (Molua 2011; Bele et al. 2013; Brown and Sonwa 2015), and thereby enhance their general well-being and adaptation potential.

Like the Anglophone and Muslim women, the incidence of malaria and waterborne diseases implies a disproportionate burden on the reproductive activities of women to nurture and care for the sick family members. However, since household farming decisions are jointly made (Gyau et al. 2014), wives have the possibility to trade-off their on-farm labor support with domestic care roles.

In summary, although female farmers across Cameroon seem to opt for relatively low resource (cost and labor) on-farm adaptation methods which can contribute to yield stabilization (Molua 2011), a variability exist among these women influenced by a combination of other socio-cultural norms and socio-economic factors. This aligns with growing literature on gender, intersectionality, and climate change which has shown that socio-cultural norms including gender relations and power dynamics which determines resource allocation and use, division of labor, and decision-making within and beyond the household influence the adaptation strategies of women. Furthermore, an intersection of other socio-economic factors jointly determines the adaptation options and choices available to women. These factors include income/wealth, educational level, place/location, marital status, class, ethnicity, group membership, access to information, technology, extension services, etc. Nevertheless, irrespective of any constraint posed by societal/sociocultural norms and other socioeconomic factors, female farmers continue to exercise their agency and adapt to changing climatic conditions in diverse ways as the circumstances present. Thus, the adaptation strategies of women within agrarian societies in Cameroon are context specific.

Caveats/Limitations of the Study

This desk study adopts a critical theorist (feminist) stance, which guides the scope of analysis and recommendations with an undeniable bias of the researcher's position. This chapter is led by a female agronomist, with lived experiences of the daily struggles Cameroonian women encounter over land and natural resources. Although the authors aim to limit the researcher's influence through self-reflective honesty and interaction with the second, more distant author, it cannot entirely erase the influence of this specific positionality, but this should not be considered a problem in this exploratory study. Furthermore, because this study relies exclusively on secondary literature (with minimal sex disaggregated data), its analysis is not exhaustive of the current realities, dynamics, and complexities of climate variability/change on agrarian livelihood systems on the ground, an in-depth field research will be needed to achieve this. Nevertheless, the conclusion and modest recommendations will shed light on the ongoing debate on intersectionality and climate change adaptation, and propose alternatives that can work in Cameroon.

Conclusion

In line with the growing debate on gender, intersectionality, and climate change, this study set out to examine the adaptation strategies of women in Cameroon using an intersectional lens. As discussed, the vulnerabilities and effects of climate change differs across the five agro-ecological zones of Cameroon with the Sudano-Sahelian zone being the most vulnerable zone. While the National Adaptation Plan of Action (NAPA) provides an institutional framework for agrarian adaptation strategies, limited coordination among sectorial ministries impedes the effective implementation of these strategies at regional and local levels. Furthermore, by classifying women as a monolithic group, the NAPA ignores the various institutional arrangements (gender norms, resource ownership, property rights etc.) and socio-economic factors (education, income, marital status, etc.) that determines the vulnerabilities and adaptation strategies of women. Thus, the national adaptation initiatives targeting women are poorly designed and seems destined to fail from its conception. Nevertheless, farmers (women inclusive) have a wide range of on-farm agricultural practices and off-farm adaptation strategies that are used to counteract the adverse effects of climate change while enhancing their livelihoods and well-being.

With land being a crucial production factor within agrarian communities, resource ownership and property rights constitute an important determinant in the adaptation choices and options of farmers especially women. However, the analysis shows that land tenure/governance across the national territory is diverse and embedded within a complex web of gendered identities, norms, roles, ethnicity, and power dynamics, for which women have to negotiate resource access with other actors. The outcome of these negotiations translates differently for women across the Muslim, Anglophone, and Francophone communities of Cameroon, and therefore influences the on-farm adaptation options and choices of female farmers in these communities.

Beyond gender and security of tenure, we have shown in this study that a range of other factors influences the adaptation strategies of female farmers across the different agroecological zones of Cameroon. These factors include decision-making power, income/wealth, level of education, access to information, access to technical assistance, group membership, and social identity/status (daughter, wife, mother, widow, or divorcee). A variability therefore exists in both the on-farm and off-farm adaptation choices and options available and adopted by women within and between communities be it in the northern semi-arid agro-ecological zone or the Southern savannah and forest agro-ecological zones. Hence, the adaptation strategies of female farmers in Cameroon is highly context specific. These findings are not unique to Cameroon, as research elsewhere have shown the intersection of socio-economic factors on climate change adaptation across Africa (see Boltana (2017) for group membership in Southern Ethiopia, Elum et al. (2017) for access to information in South Africa, Van Aelst and Holvoet (2016) for marital status and intra-household decision-making in Tanzania, etc.). Nevertheless, the study contributes to the current global debate on gender, intersectionality, and climate change by providing insights from the Cameroonian context.

Recommendations

Despite the limited sex-disaggregated and quantitative data, this study could serve as an "eye-opener" and a foundation for further research on intersectionality and climate change adaptation in Cameroon. Thus, an in-depth and contextually grounded analytical research is recommended to shed more light on the existing variabilities among female farmers within and between the different communities and agroecological zones. Nevertheless, based on this study, a few modest policy recommendations can be proposed as a way forward in enhancing the adaptation strategies of female farmers in Cameroon.

First, ensuring secured access to land for women especially within the Muslim communities can create a window of opportunity to strengthen their bargaining power and enhance their adaptation strategies. While the national land tenure reform proposed in the NAPA seems apparent as a solution, there is need to incorporate "Muslim doctrines" and make special legal provisions for Muslim women who seem to be the most disadvantaged in relations to property rights compared to their Anglophone and Francophone counterparts. Bearing in mind that security of tenure goes beyond state recognition to a matter of power struggles and social relations, sensitizing women on their "constitutional rights to property" and putting in place (or revamp existing) accompanying institutions to ensure effective implementation and gender justice will go a long way to improve the agrarian livelihood and adaptation strategies of female farmers. Also, promoting female farmer groups, cross-country exchanges and engaging with ideological debates about gender and land rights is a step in the right direction towards secured property rights for women.

Second, as education and income have shown to enhance the bargaining power and off-farm adaptation options of women within the Anglophone and Francophone communities, improving the literacy levels and socioeconomic capacity of Muslim girls and women is necessary. Beyond improved literacy levels and socio-economic capacity, education can serve as an empowerment strategy geared towards creating pathways of transforming gender norms in the long run. Thus, affirmative action through educational and vocational training scholarships could be a policy measure of the Ministry of Women Empowerment and the Family (MINPROFF) to improve overall livelihood and adaptation strategies within Muslim communities. Nevertheless, girls from rural areas, especially of enclaved areas within the Anglophone and Francophone communities, should be considered too.

Third, access to information and technical assistance has shown to be significant factors determining the on-farm agricultural adaptation practices of women. Thus, governmental effort by the Ministry of Scientific Research and Innovation (MINRESI) and Ministry of Agriculture and Rural Development (MINADER) in revamping existing meteorological services and making information and extension services more readily and widely available is crucial to enhance the adaptation decisions that women make. While priority should be given to the northern semi-arid zones with frequent droughts and floods as well as the coastal lowlands with frequent floods, storm surges, and mass movements, the other agro-ecological zones should not be neglected as the impacts of climate change in one agroecological zone inherently affects the populations in the other zones.

Fourth and last, community groups (farmer's cooperatives, self-help, "tontines," and "njangis") seem to be a viable option for peer learning, information sharing, financial assistance, and mutual support in Anglophone and Francophone communities. These groups could also serve as a transformative tool to mobilize the power of collective action in strengthening the bargaining power of women within the communities, as illustrated by Anglophone women groups in relation to access to land. The potential of these groups should be explored and supported to collectively build the adaptive capacity of women and community resilience, through initiating and training members on income-generating entrepreneurial activities, and linking these groups to microfinance institutions/banks for financial assistance through credit. Also initiating such groups within the Muslim community (if none currently exist) could be beneficial too in reducing their vulnerability and enhancing that adaptive capacity. Both MINADER and MINPROFF could spearhead such initiatives.

References

Ahidjo P (2012) Access to land and the system of land ownership in Northern Cameroon. J Knowl Manag Econ Inf Technol 2(3):1–11

Arora-Jonsson S (2011) Virtue and vulnerability: discourses on women, gender and climate change. Glob Environ Chang 21(2):744–751. https://doi.org/10.1016/j.gloenvcha.2011.01.005

Beckline M, Yujun S, Ayonghe S, Etta OL, Constantine I, Richard T (2017) Adaptation of women to climate variability in the southern slopes of the Rumpi Hills of Cameroon. Agric For Fish 5 (6):272–279. https://doi.org/10.11648/j.aff.20160506.19

Belaunde S, Cortes M, Hogstad J, Ku E, Nascimento K, Trzcinski L (2010) Land, legitimacy and governance in Cameroon. Institute for Research and Debate on Governance and Columbia University School of International and Public Affairs. 70p. Retrieved from: http://www.institut-gouvernance.org/IMG/pdf/sipa_cameroon_land_legitimacy_governance-2010.pdf

Bele MY, Tiani AM, Somorin OA, Sonwa DJ (2013) Exploring vulnerability and adaptation to climate change of communities in the forest zone of Cameroon. Clim Chang 119:875–889. https://doi.org/10.1007/s10584-013-0738-z

Boltana TF (2017) The role of women Self Help Group membership towards collective action in women's adaptive capacity building under changing climatic situations in rural Ethiopia; a case study in Kindo Koysha District; Southern Ethiopia. Masters dissertation, Institute of Development Policy, University of Antwerp, Belgium

Brown HCP, Sonwa DJ (2015) Rural local institutions and climate change adaptation in forest communities in Cameroon. Ecol Soc 20(2):6. https://doi.org/10.5751/ES-07327-200206

Carr ER, Thompson MC (2014) Gender and climate change adaptation in agrarian settings. United States Agency for International Development, Washington, DC

Cho S, Crenshaw KW, McCall L (2013) Toward a field of intersectionality studies: theory, applications, and praxis. J Women Cult Soc 38(4):785–810

Davis K (2008) Intersectionality as buzzword: a sociology of science perspective on what makes a feminist theory useful. Fem Theory 9(1):67–85. https://doi.org/10.1177/1464700108086364

Denton F (2002) Climate change vulnerability, impacts, and adaptation: why does gender matter? Gend Dev 10:10–20. https://doi.org/10.1080/13552070215903

Elum ZA, Modise DM, Marr A (2017) Farmer's perception of climate change and responsive strategies in three selected provinces of South Africa. Climate Risk Management16:246–257. https://doi.org/10.1016/j.crm.2016.11.001

Faith AN (2019) Whose land is it? Women's struggles over land in Cameroon (End-of-Module-Paper). IOB, University of Antwerp, Antwerp

Fonjong LN (2012) Equal rights but unequal power over land: rethinking the process of engendering land ownership and management in Cameroon. In: Fonjong LN (ed) Issues in women's land rights in Cameroon. Langaa RPCIG, Mankon Bamenda, pp 19–41

Fonjong L, Fombe L, Sama-Lang I (2013) The paradox of gender discrimination in land ownership and women's contribution to poverty reduction in Anglophone. GeoJournal, 78(3), 575–589. https://doi.org/10.1007/s10708-012-9452-z

Gyau A, Faith AN, Fondjem-Tita D, Ajaga N, Catacutan D (2014) Small-holder farmers' access and rights to land: the case of Njombé in the littoral region of Cameroon. Afrika Focus, Special Agroforestry Issue 27(3):23–39

Hankivsky O (2014) Intersectionality 101. Institute for Intersectionality Research and Policy, Simon Fraser University, Vancouver

Innocent NM, Bitondo D, Azibo BR (2016) Climate variability and change in the Bamenda highlands of North Western Cameroon: perceptions, impacts and coping mechanisms. Br J Appl Sci Technol 12(5):1–18. https://doi.org/10.9734/BJAST/2016/21818

Kaijser A, Kronsell A (2013) Climate change through the lens of intersectionality. Environ Polit 0 (0):1–17. https://doi.org/10.1080/09644016.2013.835203

Kings AE (2017) Intersectionality and the changing face of ecofeminism. Ethics Environ 22 (1):63–87. https://doi.org/10.2979/ethicsenviro.22.1.04

MINEPDED – Ministère de l'Environnement, de la Protection de la Nature et du Développement Durable (2015) Plan National d'Adaptation aux Changements Climatiques du Cameroun. Yaoundé, Cameroon: 138 pages + Annexes

Molua EL (2009) Accommodation of climate change in coastal areas of Cameroon: selection of household-level protection options. Mitig Adapt Strateg Glob Chang 14:721–735. https://doi.org/10.1007/s11027-009-9194-5

Molua EL (2011) Farm income, gender differentials and climate risk in Cameroon: typology of male and female adaptation options across agroecologies. Sustain Sci 6:21–35. https://doi.org/10.1007/s11625-010-0123-z

Molua EL (2012) Gendered response and risk-coping capacity to climate variability for sustained food security in Northern Cameroon. Int J Clim Change Strat Manag 4(3):277–307. https://doi.org/10.1108/17568691211248739

Ngondjeb YD (2013) Agriculture and climate change in Cameroon: an assessment of impacts and adaptation options. Afr J Sci Technol Innov Dev 5(1):85–94. https://doi.org/10.1080/20421338.2013.782151

Njoh AJ, Ananga EO, Anchang JY, Ayuk-Etang EMN, Akiwumi FA (2017) Africa's triple heritage, land commodification and women's access to land: lessons from Cameroon, Kenya and Sierra Leone. J Asian Afr Stud 52(6):760–779. https://doi.org/10.1177/0021909615612121

Osborne N (2013) Intersectionality and kyriarchy: a framework for approaching power and social justice in planning and climate change adaptation. Plan Theory XX(X):1–22. https://doi.org/10.1177/1473095213516443

Pumenta NV (2014) The impact of climate change on food security and health in Northern Cameroon. In: Keyes CB, Lucero OC (eds) New developments in global warming research; climate change and its causes, effects and prediction. Nova Science Publishers, New York, pp 1–55

Pumenta NV (2017) When 'property cannot own property': women's lack of property rights in Cameroon. Afr J Econ Sust Dev 6(1):67–85

Sonwa DJ, Somorin OA, Jum C, Bele MY, Nkem JN (2012) Vulnerability, forest-related sectors and climate change adaptation: the case of Cameroon. Forest Policy Econ 23:1–9. https://doi.org/10.1016/j.forpol.2012.06.009

Tingem M, Rivington M (2009) Adaptation for crop agriculture to climate change in Cameroon: turning on the heat. Mitig Adapt Strateg Glob Chang 14:153–168. https://doi.org/10.1007/s11027-008-9156-3

UICN-PC (2015) Stratégie de Prise en Compte des Femmes dans le Processus REDD+ au Cameroun. UICN (Union internationale pour la conservation de la nature et de ses ressources), Yaoundé, Cameroun: 73 pages + Annexes

Van Aelst K, Holvoet N (2016) Intersections of gender and marital status in accessing climate change adaptation: evidence from rural Tanzania. World Dev 79:40–50. https://doi.org/10.1016/j.worlddev.2015.11.003

Van den Berg A (1996) Land right, marriage left: women's management of insecurity in North Cameroon. Research School CNWS, Leiden University, Leiden

Yengoh GT, Tchuinte A, Armah FA, Odoi JO (2010) Impact of prolonged rainy seasons on food crop production in Cameroon. Mitig Adapt Strateg Glob Chang 15:825–841. https://doi.org/10.1007/s11027-010-9241-2

11

Sustainable Urban Drainage Practices and their Effects on Aquifer Recharge

Getrude Gichuhi and Stephen Gitahi

Contents

Abstract

Between 1994 and 2006, an 18% increase of freshwater flow into the earth's ocean was recorded, as well as extreme weather events such as prolonged drought and intense floods. Following this period was an era of increased evaporation from oceans and seas, which heightened global warming in Africa. This chapter proposes the use of man-made aquifers recharge processes as methods of draining water into the soil before the runoff water reaches water bodies. **Source control** involves controlling the volume of water entering drainage systems or rivers by cutting off runoff water through storing for reuse or evapotranspiration as seen in green roofs. **Pre-treatment** is the use of trenches to filter and remove

G. Gichuhi (✉) · S. Gitahi
Department of Research and Innovation, Strathmore University, Nairobi, Kenya
e-mail: ggichuhi@strathmore.edu; stephen.gitahi716@strathmore.edu

contaminants from water before getting to water bodies. **Retention systems** on the other hand is controlling the rate at which water is discharged to waterways by providing water storage areas such as ponds, water retention areas, etc., while **Infiltration Systems** are areas that allow natural soaking of stormwater runoff to the ground naturally recharging the water table. The proposed methods will see replenishing of the water table, a great leap in the efforts of curbing global warming. This practice can easily be adopted by both individuals and government as we build more and more buildings creating a balance between the need for human settlement and the natural way of water replenishing itself. The methods do not introduce extra costs to an already existing budget. In some cases, the methods help to reduce the costs of projects especially in urban areas. Africa which hosts many of the growing countries sees and will continue to experience surges in urbanization. For such, these methods presented in this topic will be, if implemented, a best method to solve the urban drainage problems before this even occurs.

Keywords

Aquifer · Hygiene and sanitation · Groundwater recharge · Global warming · Urban water balance · Universal water security · Topographical land

Introduction

Since 1990, water rise in the sea and oceans has been recorded and is now more visible than any other time in human history (The Ocean Portal Team 2018). The highest attributed factor has been melting of glaciers on ice topped mountains and concrete constructions. This ice melts and flows as runoff water into small water collection points such as ponds and swamps to bigger water bodies such as rivers, lakes and oceans increasing water flow into water bodies. All these changes result in an increased global water cycle where water moves cycles through evaporation, precipitation, and runoff, an aftermath of increased rate of global warming.

Methods of mitigating the negative effects of climate change in Africa have become a big issue. Individual companies and governments in Africa are trying to implement resilient ways that adapt to the effects of climate change. African countries are prone to the effects of climate change especially because of the higher levels of industrialization. On the other hand, developed countries that have implemented ways of harnessing water and avoiding runoff water directly to catchment areas at all costs is something Africa has not been able to fully accomplish (The United Nations-Water 2015). According to a report by UNEP, by 2020, between 75 million to around 200 million people in the African continent are projected to experience water stress due to climate change (UN Environment Programme 2020). In this twenty-first century, most countries in Africa are becoming urbanized; modern housing, infrastructure, access to water, and implementation of technology has become common (Ogwu 2019). Narrowing down to East Africa, Kenya is the most urbanized country and then followed by Rwanda. Cities are evolving and the idea of implementing sustainable urban drainage is still an issue as they improve the topography of cities.

Kenya's fastest growing cities include Nairobi, Mombasa, Thika, Machakos, Nakuru, Malindi, and Kisumu. Untapped but identified as potentially stable upcoming cities include, Tatu city and Konza city. This research identifies these two areas as the best place to promote sustainable urban drainage (SUD) systems and aquifer recharge is implemented. Other cities will follow the example set by these two as the benefits come to play. The total population for Kenya currently is at 54 million, according to the 2019 census; Nairobi, the capital city has just over four million people (Kenya Data Portal 2019). Water Security in a country is one of the major issues facing each country worldwide. According to Sustainable Water Partnership, water security is the adaptive capacity to safeguard the sustainable availability of, access to, and safe use of an adequate, reliable, and resilient quantity and quality of water for health, livelihoods, ecosystems, and productive economies (Viala 2016). UN Water defines water security as the ability of a community to protect their sustainable ways of accessing quality water (UN Water 2013). This is very vital especially to areas that have a huge population. Insufficient water in some areas is caused by drying up of the aquifers, while some areas are faced with excess water either due to flash floods, or rainwater runoff, which in turn leads to less aquifer recharge. These imbalances are a clear result of insufficient aquifer recharge. With no improved strategies to better ways of harnessing water, especially with the current climate projections showing irregular weather patterns, aridity of these cities is prone to increase at a very high rate. A case point is Two Rivers Phase II project, a modern residential center hosting a mall in Nairobi and Kiambu Counties in Kenya, which if the runoff water is not taken care of, the drying up of aquifers might kick off. Water in Two Rivers in which the development seats are likely to rise. Coastal areas are also facing similar issues. A case of concern there is salty water finding its way into the water table recharging the aquifers and thus making the water for consumption saline.

Rural-Urban transformation has improved the topography of most countries in Africa. Although, this has culminated in the improvement of settlement areas as well as the economic status of residents, it has heightened developmental challenges such as making water runoff move to a collection base settled in one area, which either flows to other big catchment areas or slowly evaporates. This chapter aims to put more emphasis on ensuring that sustainable urban drainage systems are implemented in a bid to see that boreholes that source water from aquifers are managed responsibly ensuring water availability and safety. Africa has not yet formulated the policy of adapting the best practices of sustainable urban drainage systems and natural aquifer recharge methods. Aridity of land, increase of saline water, drying up of aquifers, and increased water wastage, especially with increased upcoming constructions, are an indication of the state of globalization and effects of climate change in a country. Sustainable Urban Drainage Systems are drainage networks that act as collection points, transport, treat, retain, infiltrate and drain runoff water in sustainable ways that are beneficial for living things and the world in general (Hidrologia

Sostenible n.d.). Given this background, the aim of this chapter is to review the status of sustainable urban drainage and Aquifer recharge within African context, highlighting some technological advances to be used in the current scientific and topographical challenges of climate change.

Related Works and Existing Gaps

This chapter uses referencing points, of developed countries, and how they implemented the SUDs, which affects the natural recharge of Aquifers. According to Matos (2003), drainage systems are not necessarily a rule when it comes to construction of infrastructures, and runoff water would cause flooding of the roads or clogging of the sewers. On the other hand, Tim (2008) speaks of the Christian era, where drainage systems constructed by the Romans in the BC to AD are still functioning till date. Introduction of sanitation began in Italy after an incident of septic tank sinking due to floods and the country was forced to eliminate flooding areas and adapt underground channels, which was a wake-up call to other European countries in adapting sanitation systems (Poleto and Tassi 2012).

The use of green gardens in developed countries that substitute farming due to construction of buildings covering the whole land makes it necessary to adapt previous ways of MAR. An important agenda highlighted in this chapter is to understand how policies are followed to the latter for avoidance of implementing after math eqs. A study conducted in the United Kingdom regarding their technical, legal, and planning methods of control for urban runoff water in their country was done to ensure countries that do not have policies created to govern engineers before starting projects (Woods et al. 2015). Governance of Aquifers needs to be implemented. There are three aspects done in developed countries as to the governance of aquifers, management of discharge of water, management of recharge, and substituting alternative water sources.

Management of discharge of water from the aquifers currently is important as drawings from these sources are at a high rate, especially due to the domestic use of the water. Some countries use water mainly for commercial use while others use water to pump for electricity, irrigation and domestic use as well. As depletion of aquifers becomes more and more real, countries are forced to dig deeper boreholes with extra costs to incur to reach water levels (Dillo et al. 2011). These scenarios put more weight on the need to ensure recharge of the aquifers is managed well. Management of Recharge to the aquifers can be implemented both artificially and naturally. If 95% of the land is covered with soil and grass then the possibility of water sipping into the soil is most likely to happen faster reducing the amount of water that flows into water bodies. With increased urbanization, more innovative methods of recharging of aquifers are practiced by directing water to catchment bodies, and using filtration methods that enable water to be soaked into the soil and thus refilling the aquifers. Other alternative ways of using water sources include using rainwater and customizing it for consumption use.

Due to urbanization, most countries experience a lot of runoff water, which therefore needs a number of managements of aquifer recharge procedures through identifying a number of solutions that gets back to the aquifers. These processes have been practiced before, and there is a need to introduce them in developing countries especially due to the increase of urbanization in these same countries. Use of permeable floors could be one of the implementing ways.

There are other substituting alternative water sources, which are currently used all over the world. Some authors have done research on how to tap water and collect it in a catchment where one can fetch for domestic use. Other water sources enhancements include channeling rain water through use of gutters, tapping water from a river into a house. More methods have been adapted in today's world to ensure that they have sources of water. Some negative points have been brought forward, such as some of these ways being harmful to the humans and the animals. A case study was carried out in Kilifi County in Kenya where the test was to find out the presence of arsenic in water and the pH of the same water. According to the World Health Organization (2003), pH of consumable water should be between 6.5 and 7.5.

A study research was done on SUDs in low-income countries in Africa outlining the disadvantages that runoff water instills such as soil erosion and siltation (Reed 2004). Some special cases have been raised, for instance, constructing riparian lands and coastal areas. These areas have a land mass that is not stable with fault lines that aid in the recharge of aquifers experience. Kenya has had cases of demolition of buildings due these buildings being on riparian lands, this was a threat not only to the SUDs not being observed but also to the lives of people who were meant to live and operate in such areas. Other concerns raised, is the groundwater overexploitation in the coastal regions and these buildings being on riparian lands. Groundwater overexploitation in the coastal regions leads to intrusion of salty water (Geiger 2020). This chapter will help us understand the options that a country has in case it is forced to build on riparian lands.

Some practices recommended in this chapter exist in developed countries and can be very effective in African countries in understanding climate change and how to deal with the effects of those changes. There are challenges facing the Kenyan topography such as cases of constructions being done on riparian lands. Such areas are not stable and could result in the collapse of such buildings. It has resulted in the government demolishing such houses and prosecuting responsible constructors for not adhering to the rules and policies that guide the said industry. Coastal lands on the other hand have cases of saline water recharging the aquifers, making it difficult for consumption especially to humans. The best result in these situations is to ensure filtration methods are used.

Policy making for engineers and owners is one of the proposals for governments to implement. Adaptive measures of some proposed solutions are suggested in order to ensure smooth implementation of SUDs and aquifer recharge as well. These include filtration methods and systems of reducing salinity and the productive measures of constructing buildings in riparian lands. Resolving these issues with the proposed methods will see an increase in the amount of water that goes back to the water table, replenishing it, as opposed to flowing into the ocean and seas. On a concrete surface, 95% of water on the surface runs off, only 5% gets absorbed into the soil. With natural aquifers taken into account in the landscape planning and designing, the 95% of the runoff water will be adsorbed to soil, resulting in only 5% of the water being drained to the oceans and seas. This mode of replenishing the water table ensures that wells and boreholes dug in arid and semi-arid areas do not dry up fast. The benefits of this extend to what such areas can do in their economic ventures and other activities made capable by the presence of stable water supply.

Aquifers

When Water soaks into the soils from the rivers, melted ices, and rain, it is concentrated in one area that forms its own settlement, which is referred to as an aquifer. This settlement of water is formed in different bowl-like areas that form a well and are all in different areas with different levels. A number of aquifers formed in one area are referred to as the water table level forming the zone underneath the ground. There are two types of aquifers, the confined aquifers and the unconfined aquifers. Confined aquifers are one that is way below almost at the bedrock of the rocks and formed by permeable rocks due to the pressure of non-permeable rocks. However, getting to the confined areas at times seems difficult and collapsing of the rocks could hinder reaching the aquifers when drilling boreholes. Unconfined aquifers are the ones that are directly refilled with the water soaking directly from the surface of rivers or when in a swampy area or irrigation is taking place. The unconfined aquifers are the ones that rise or drops depending on the amount of water stored.

Aquifers in rural areas and with a water source around are less tapped as compared to an arid area. In semi-arid areas where these boreholes are drilled with the changing weather conditions and less water source reservoirs, humans turned to aquifers, through drilling boreholes in pursuit of water. Sometimes the water turns to be saline, depending on the locality. For instance, coastal areas tend to have more saline aquifers as compared to fertile moisture soil with cold weather conditions. As urbanization in coastal areas increases, the drilling of boreholes is increasingly at a higher rate. This in turn, due to unconfined aquifers being empty, is refilled with saline water from the ocean. This is reusable, however not as drinkable water for humans nor animals as well.

Consistent use of these aquifers without recharges is only sustainable for a few decades. According to Kansas States University in Manhattan, a 4-year research of the Ogallala Aquifer was conducted to identify the use of aquifers for irrigation and consumption use in Kansa, they found out that continued use with no recharge could cause depletion of food production in that area (Chow 2013). Glaciers melting sorted that situation for a while with the chances of aquifer usage to be in survival for 50 years, after it only used up to 30% of the water.

Glacier melts in areas whose topographical land is associated with mountains and areas that are cold and can experience snow. Africa, has very minimal of such topography and therefore, a few proposed solutions on ways to manage the discharge of the water from the aquifers, how to recharge them and alternative ways of water use to avoid depletion of the groundwater.

Research Framework

Looking at the sustainable urban drainage practices that are done to the developed countries and the effects of aquifer recharge, there is an understanding that SUDs consideration will be established in Africa. With a different topography, there is a need for an analysis of the geographical area to determine any potential trends that are made or can be made. The analysis done is to determine a number of things, the risk factors involved in case an SUD is proposed in that area, and the number of SUDs to be implemented in that area, and the demographics as well. After analysis, the topographic area will help in determining the type of qualitative criteria to be used for the SUDs.

Criteria	Relevance	Category
Guidance documents	These are the documents to be used as the implementation process for the SUDs in Africa. The use of these documents either required by the government or voluntarily offered to the government by the implementing partner shows the extent to which the SUDs would have effects on the aquifer recharge. These guidance documents can be used as practices that any organization or individual looking into SUDs should have, and also used as ways to overcome over implementation of SUDs in one area	Guidance documents not specified to SUDs but to construction – Policy constructions documents Standard government guidance for a country from the National Construction Authority Relevant government planning policy statements concerned with hydrology, and water as natural resource, e.g., a) delivering sustainable development policy b) Planning and pollution control policy c) Disaster and risk management policy – This looks deeper into drought and floods as ways of managing disasters and risks in a county
National Construction Authority Consultation	In order to ensure involvement, in the case of Kenya, it is important to ensure the National Construction Authority is involved for assessing the area, and the risk factors that could be involved	Consulting the National Construction Authority of a country, for location intelligence analysis, through the use of documents, records of previous assessments, reports, and telephone and email conversations.
Constrains	Financial constraints in case of a collapse of a borehole drilled or a pond created are some of the constraints involved when it comes to implementation of SUDs	Other constraints could be through desktop and site survey CLA not specified and not included in analysis due to no knowledge of contamination at site. 1 – CLA through desktop survey only. 2 – Desktop and site survey

(continued)

Criteria	Relevance	Category
SUDs type	This chapter categorizes SUDs into substituting alternative water sources control, Management of Aquifer recharge, and Management of Discharge of water	**Substituting alternative water sources control** (a) Green roofs (b) Rainwater harvest (c) Redirecting water retention systems **Management of Aquifer Recharge** (a) Infiltration systems (b) Treating municipal wastewater (c) Treating storm water (d) Treating irrigation water **Management of Discharge of water** (a) Retention systems (b) Wetlands
Numbers	Identifies current practice and baseline data	Types of SUDs to be used in that area
Size	Identifies current practice and baseline data	How large or small will the SUDs be used
Efficiency	Identifies current practice and baseline data	How sustainable is the SUD used
Maintenance	Identifies current practice, baseline data, potential barriers, ways to overcome them, and trends	How frequent should it be maintained and treated
Good practices	Africa as a developing continent, most of the topographical area is improving day by day, which makes runoff water become a potential problem especially to rural areas. Therefore, different mitigation process of water should be implemented in the policy making and as part of construction when it comes to buildings and roads	Use of best practices. Evidence of National Construction Environmental Management Plan or equivalent.

Proposed Solutions

On the account that there is uncertainty when it comes to climate change and the shortage of water especially in Africa, adaptation strategies on water usage and refilling of aquifers should be prioritized. Clear projections on water demands in the future are merely determined by the climate change, sustainability imbalances between freshwater, and demand for it (Taylor et al. 2009). There is a huge difference of climate change association in developed countries as compared to the developing countries. In addition, the fact that most countries in Africa are in the sub-Saharan desert, the usage of water is more likely to be at a higher demand and also depletes at a higher rate.

Although, some of these solutions are implemented in other countries, recommending them to African countries, suits best especially during this time of globalization. African countries are known to be experiencing different kinds of weather conditions, as compared to developed countries. The Sub-Saharan desert covers most parts of Africa. Solutions proposed in this chapter targets what African countries would experience when it comes to recharge of aquifers as well as substitute water sources. Runoff water and the impact it has on climate change has the need of ensuring that the community is able to implement SUDs and redirect water to aquifers. This chapter discusses the methods to be used in African Countries in the management of discharge of water, management of recharge, and substituting alternative water sources.

Substituting Alternative Water Source Control

It is clear that using water from Aquifers is not only the one area that humans should depend on if it is for consumption at home or commercial level. There are other means of attaining water with source control implemented. Source control is a way of managing the volume of water entering drainage systems or catchment areas by cutting off runoff water either to be redirected into another source stored for reuse or evapotranspiration or to absorb into the soil as a way of recharging the aquifer (Geiger 2020). Mostly runoff water will be from rainwater, flash floods, or melting of glaciers. Besides, source control can be exercised when the water is settled in one area with no use of it. Wet areas are an example of how source control can be used.

A country's topography has around four features that affect how humans will cultivate and settle around that area (National parks Association n.d.). These are the landforms, which include the hills, valleys, gullies, and mountains, etc. These places act as animals settling places but as a tourist attraction to humans, which affects the economic status of a country. A water source is another feature of land topography, which mostly includes the rivers, swamps, coastal oceans, etc., which have a great impact on affecting our sources of water. Vegetation includes the national parks, the farmlands, etc., acting as the main source of food to both humans and animals. Manmade topographic features, which fall under the category that most humans ensure that they plough and invest in, comprises of buildings, roads, property boundaries etc.

The four features of land topography have a huge effect on aquifer recharge; the recharging of the aquifers depends on how much is interrupted. For instance, if the sources of water, rivers, swamps, oceans are polluted, the same water used to recharge the aquifers will reach the table water polluted which becomes harmful to human beings and animal consumption. On the other hand, if manmade features of the topography are not structured in a way that allows runoff water to seep back to the ground the possibility of the aquifer to recharge is of bare minimal. The untapped areas such as landforms and vegetation are most likely to have the water table to be fully recharged or at least for use through soil moisture and suitable for plantations.

Green Roofs

The level of a city becoming smart has better ways of ensuring that it implements source control mechanisms. Green roofs are a way of turning barren landscapes into a living network of gardens (Willem 2005). It is a type of urban agriculture and a way of saving space and at the same time living expenses are reduced. Green roofs have a number of advantages such as:

(a) Reduction of runoff water – one of the main aspects that African Countries are trying to fight
(b) Act as a temperature buffer for the home. The plants combat the natural heat and coolness from the atmosphere
(c) Economical potential of cultivating food from the green roofs

Green roofs comprise a multilayered system that covers the roof of a building with vegetable landscaping. The roof is likely to consist of an impermeable layer, a growing medium, and a drainage layer. The impermeable layer aids in the survival of the plants by increasing the retention time of water in the substrate layer thus increasing the soil moisture. The Substrate layer or the growing medium layer retains the water and uses the right amount of organic and mineral proportions. The drainage system is the most important part in setting up the Green Garden. There are different types of drainage systems; however, the ideal one is that which provides good circulation and evacuation while at the same time offering water supply to the crops. In today's modern green garden drainage system, they use rigid and open mesh structures plastic sheets (Critical Concrete 2018).

Although, not to say that green roofs should be used in places where aquifer recharge is a priority, it only states that it can be used as an alternative source of acquiring water for consumption use and irrigation use as well. The implementation of Green roofs especially in urban areas of a country ensures hydrologic-hydraulic invariances in that area. This is one of the main areas that this chapter tries to focus.

Rainwater Harvest

Rainwater harvesting refers to different methods used to channel rainwater into an area where it can be used for other purposes (Black et al. 2012). There are various types of rain water harvesting systems used in different countries, which depends on the cost, size, and the complexity of the technology involved. However, the main aim is to ensure utilization of rainfall to be used as an alternative source of water both domestically and commercially. The beauty of it is that it can be used in a small or large industry or even at the homestead level. With the current climate changes, using resources to ensure the survival and usage of clean water in the future has been one of the ways to fight drought.

The collection of rainwater from rooftops, or other surfaces or even open space into a catchment area is known as rainwater harvesting. This harvested rainwater could be used commercially, domestically, or for farming purposes (irrigation). On the other hand, harvesting rainwater could act as an advantage by providing stormwater management especially in urban areas thus leading to detention pools, which are used to recharge aquifers. Slow runoff water has a higher chance of soaking into the ground, which has other advantages such as reducing contamination and avoidance of soil saturation.

The types of rainwater harvesting include:

1. Domestic rainwater harvesting. There are two types of harvesting systems used especially in developed countries: gravity systems and pump feed systems. Gravity systems are positioned at the top of a building, which uses gravity to propel water to different parts of the house. Examples are the use of gutters to direct water to the tanks. Pump feed systems on the other hand, is when water is pumped upward toward different water holding objects or tanks to be used for domestic use. Drilling of boreholes is an example of pump feed systems.
2. Commercial rainwater harvesting. There are different types of harvesting systems that are used in large farms. The most common one is water butts, which is one of the common harvesting systems used in the UK. Water butts are barrel-like or bin-like tanks that collect water from drain pipes to store water that could be reused for farm use (The Renewable Energy Hub UK 2018).

Although, at times the collected water may not be user friendly, the level of treatment when it comes to harvested rainwater depends on the use required and the dimensions of the size of a home or industry. Water required for Aquifer Recharge will definitely undergo a strict way of treatment as compared to one used for irrigation purposes. Also, water required for human consumption will require strict treatment unlike farm use or irrigation; some TDS would be left in the water in order to give nutrients to the plants.

Redirecting Water in Retention Systems

Urbanization in today's world is more of covering all land with cement, leaving only the rural with soil. This significantly increases the runoff water to drainage systems, which are redirected to the rivers or other water catchments. In Rural areas the rate at which water soaks back to the soil and absorbed back to the soil is at around 95% as compared to an urban area, in which the water soaks only 5% that at least lands on soil, leaving only 95% running off. Urbanization in a retro sense has its effects on affecting water that flows. However, there are precautions that engineers can take when constructing buildings and roads.

In cases of riparian lands, as stated earlier in the two rivers case, policies and design ideas ought to be structured to ensure that water is not blocked. Water flowing rates, its magnitude, frequency, and duration should be part of the measures that are included when constructing in riparian lands. In addition, these features can determine at what rate and amount it will soak into the ground and down to the aquifers. Therefore, channelling the water and either use it is another means of source of water for commercial use or constructing measurable ways of ensuring that there is no runoff water. For swampy areas, at times the ground is too full to absorb more water and will result in flooding of homes, in this case there should be a channel that redirects where this water needs to be absorbed back to the ground in a land that is a bit semi-arid.

Riparian land has been managed before using three practical methods, the Proper Functioning Condition (PFC), the Hydrogeomorphic Approach (HGM), and the index of Biological Integrity (IBI). In spite of the advancements in technology, African Countries have not yet implemented such strategies. PFC is a qualitative approach and depends on the knowledge and judgment of a team of experts. HGM and IBI on the other hand are based on quantitative data gathered and analyzed from unaltered to degraded sites prior to assessor's involvement (National Research Council 2002).

Management of Aquifer Recharge

Infiltration Systems

In the intent of refilling deep aquifers, a process involved, happens at a gradual rate through the unsaturated zone all the way to the aquifers. Although, there are underlying factors that depend whether the aquifer is suited in an area where it was used repeatedly then refilling it might take ages to do so. The creation of infiltration systems will look into a number of risk assessments, which include, land cover, soil characteristics, soil saturation, slope of the land, and evapotranspiration (Alley et al. 1999).

Dependency on water and the rate at which the water is used while it depletes in the aquifers depends on the population of a country; it is not only used by humans but also animals. Controlled ways of recharging the aquifers need to be implemented in order to ensure safety for the consumption of water. There are two ways of recharge, artificial recharge and Natural recharge. Artificial recharge is the process by which excess surface water is directed into the ground either by spreading on the surface, using recharge wells or altering natural conditions to increase infiltration. It is a way of storing water in case of a shortage.

There are a number of ways recharged water can be used. It can be redirected to semi-arid areas and arid areas for consumption and irrigation schemes. In coastal areas, with the use of artificial recharge, since it is fresh, this could be used as a control mechanism in avoiding seawater or saline water intrusion to the aquifers. There are three types of infiltration systems for aquifer recharge as a source of water, Treated Municipal Wastewater, Storm-water runoff, and irrigation water. In order to implement infiltration systems, risk assessments should be conducted by the government to ensure safety measures are in order. Infiltration systems largely influence ground stability and water quality as the latter flows to water streams. Examples of such systems are infiltration trenches, basins, and permeable paving.

Treating Municipal Wastewater

For the Municipal water of a country ensures that it treats and input infiltration systems there are a number of things they would consider. The quantity and quality of wastewater in an area, the commercial and industrial establishment off an area, and the condition of the sewer system. The municipality has to analyze the wastewater at their disposal and then run the ingredients they are to concentrate and

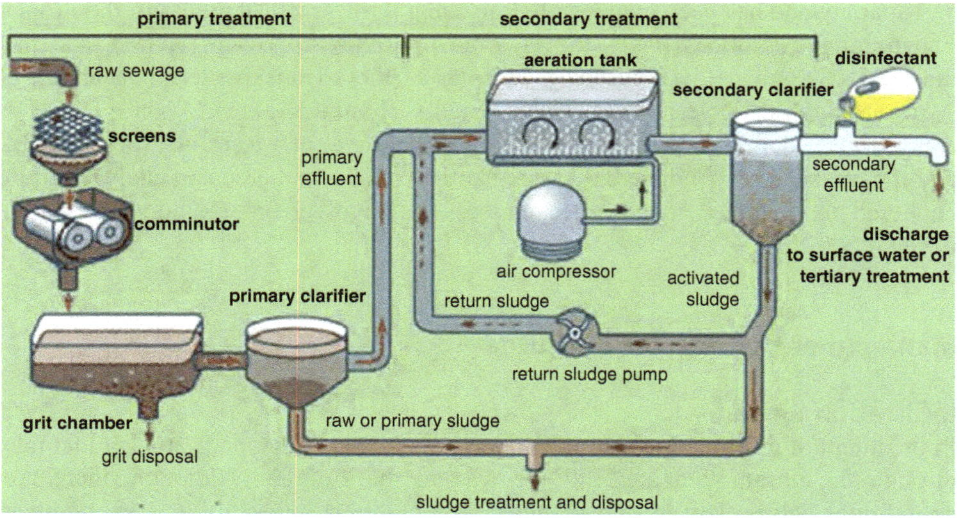

Fig. 1 Figure of the treatment process of Municipal Wastewater Infiltration process. (Eschool 2018).

dissolve. Untreated wastewater contains an amount of microorganisms' concentration. These factors include originality of the water used, the general health of the contribution population (are they in a slum area or in an urban area), and the infectious agents that survived outside the hosts under a variety of environmental conditions (The viruses, in a group become resistant to the environment that kills bacteria). Wastewater treatments are classified in primary, secondary and advanced level (Fig. 1).

The wastewater is first pumped to the system, and as a primary treatment, it is screened, then taken through to the grit chamber where the solid and huge particles are released, and finally to the secondary stage through the clarifier. The secondary treatment stage incorporates aeration process, which after takes the advanced treatment stage, which is the final stage before the water is directed to the aquifers for refilling or tanks for industrial or household use.

a) **Primary treatment of municipal wastewater**

This is the first step that includes screening of the state of the wastewater and grit removal. Once the water is screened, large solid particles are removed to avoid any interference at a late stage of treatment. Grit chambers are created as part of the infiltration systems to remove sand, seeds, glasses, eggshells, etc. There are other preliminary treatment operations such as the flocculation, odor control, chemical treatment, and pre-aeration. Primary treatment has very little effect when it comes to the removal of other biological species, which means it does not remove viruses at this stage

b) **Secondary treatment of Municipal wastewater**

This step is intended to remove soluble and colloidal biodegradable organic matter and other suspended solids (SS). This treatment involves an aerobic biological process whereby microorganisms oxidize organic matter in the wastewater.

c) **Advanced treatment of Municipal wastewater**

This is the third stage of removal of nutrients, for instance, phosphorus and nitrogen and SS. This is the physical, chemical, or biological treatment process used to accomplish a degree of treatment. It removes SS and dissolved substances either organic or inorganic in nature. This is necessary especially if there is a direct injection into the aquifers for recharge. The major processes in this stage include, coagulation-sedimentation, filtration, nitrification, denitrification, phosphorus removal, carbon adsorption, and reverse osmosis.

Stormwater Runoff

Stormwater runoff is one of the main contributors when it comes to aquifer recharge. However, pollution of the stormwater may contaminate the aquifers as well once they reach them, which needs treatment before they reach the aquifers. It is erratic in quantity and the timing as well could be when flash floods happen, or when a heavy rain has occurred. According to Larry, he tends to believe that season has a major impact when it comes to storm-water runoff quantity and the base flow (Mays 2001). The topography of an area has a major impact when it comes to stormwater pollution concentration, which tends to filter the pollutants. Impervious surfaces on driveways, parking areas, gutter drainage systems, and roads reduce infiltration of runoff and runoff to the ground.

Detention ponds are constructed for specifically recharging groundwater through the bottom of the pond. The water filters slowly through the ponds and to the aquifer, they are created deep enough to prevent water plant growth but shallow enough to prevent anaerobic conditions formed at the bottom. Inspection is necessary occasionally to ensure that treated water is what recharges the aquifers.

As a concept of salt evaporation ponds, water is directed into a pond where it is stored while it slowly soaks into the ground to the aquifers. In Kenya, Kitui District, Sand dams were created in order to store water, although this was only done in large-scale farms. A sand dam is constructed as a small dam, in and onto a riverbed of a seasonal river, where it is accumulated with sand. The sand layer acts as a filter for removing contamination and enabling water to reach the aquifers and recharge them. With areas that frequently receive huge rainfalls, there are chances of groundwater levels being high considering the use of multiple sand dams.

Subsurface dams is another way of ensuring an aquifer recharge is exercised. They are created in semi-arid areas at the seasonal riverbanks, by digging a trench to the bedrock or other impervious layer. This is something that is exercised in Brazil and can be implemented in Kenya especially in semi-arid counties, such as Kilifi county. To prevent particles soaking into the ground, use of canvas when creating the trench is necessary to act as filters, which could puncture the fabric used.

Treating Storm-Water

Hydrogeologists ensure that they calculate the right places for constructions on ponds and measure input when it comes to treating the water to be used to refill

Fig. 2 Figure of Filtration process for stormwater runoff (Fluence News Team 2017).

the aquifers, as well as calculating the right places that have aquifers that have enough water to be used. Sedimentation is the process of allowing suspended particles in water to settle out of the suspension under the effect of gravity. These particles become sludge, which is assisted by mechanical means that cause it to thicken (Bache and Gregory 2007).

Although removal of particles can vary from one version to the other, the variations of storm-water filtration may include vegetation filtration, drainage systems, and impervious pavements. The settlement of particles in ponds varies in different ponds in different areas. Therefore, when it comes to filtration either natural cause of treatment could occur when it comes to biodegradation, biotransformation, and bioaccumulation or the sedimentation process (Fig. 2).

Irrigation Water

Irrigation is considered one of the most important roles in food production in pursuit of food security of a country. Although exploitation of aquifer water for irrigation, and consumption use, particularly during dry season, has caused the groundwater level to decline in semi-arid and arid areas, there is no direct recharge of shallow aquifers. With the increasing use of groundwater, the rise of hydrogeological and climate change problems are becoming major setbacks of certain countries. However, Irrigation Return Flow (IRF) is a promising way of recharging the aquifers for better sustainability in the future.

Irrigation return flow is the excess drainage water collected from irrigated farmland, which is considered as not evapotranspiration or evacuated by surface water and is returned to the aquifer (Dewandel et al. 2008). Depending on the types of crops, there are likely to be a number of agricultural chemicals applied, with a varying range of quantity input to enable growth and avoid pests. Irrigation is mostly used in semi-arid and arid areas; this increases the salinity of the IRF to three times that of normal water.

Treatment

Numerous things are involved in the irrigation return flow water that includes TDS, nitrogen and phosphorous compounds, pesticides residues, and, at times, metallic substances. There are different kinds of technologies that have been implemented in different countries, especially with advanced technology, to ensure removal of such pollutants.

Reverse osmosis system is one of the technologies used; it is a process that uses a permeable membrane to remove TDS. The first step of reverse osmosis is the sediment filter. It is a polypropylene sediment pre-filter that removes SS from the water. Then the carbon granule filter is activated, which removes the chlorine and other organic substances, unpleasant color, tastes, and odors from the sediment water. The third stage is to activate the carbon block filter, which removes the remaining chlorine and organic substance without the carbon fines. The RO Membrane is the fourth stage that rejects the wide spectrum of impurities including bacteria and viruses. Finally, the polishing resin carbon filter dissolves gases to ensure freshness and natural tasting clean water (Hydrolink Technologies 2014).

Farm-led management is another implementation that most developed countries are undertaking. This involves training farmers on crop selection in relation to soils, methods of increasing yields, community-based pond water monitoring, and evaluation and knowledge management on aquifers. Besides that, training the farmers on the consequences of excessive use and implementation of well-directed aquifer protection measures has helped farmers to understand the quantity of water usage, fertilizers, and wastes to use.

Management of Discharge of Water

When water moves out of the saturated area or ground to the surface through springs or seeps, it is referred to as groundwater discharge. However, at times, groundwater discharge can be controlled, where water is pumped to a retaining tank or catchment basin. To ensure that water is used at a manageable rate, to avoid depletion of groundwater, control has to be instilled to ensure that this happens. There are various ways to ensure this happens, through creation of retention systems, existence of wetlands, and others.

Retention Systems

These are water systems meant to control the rate at which water is discharged to waterways by providing water storage areas such as ponds and small-scale basins. Retention systems are systems used to store water that is likely to be used for a longer period unlike detention systems that only store water for a short period and thereafter redirect filtered water back to the ground to refill the aquifers.

A case in point is, as the rainfall season approaches, seasonal rivers tend to overflow and at times cause flood in certain areas especially ones that are arid. In this case, building of retention systems to pump water from the river to the systems could help in the use of the water during dry seasons.

Wetlands

Another way that discharge of groundwater happens is through the existence of wetland areas. The riparian lands as well as the swampy areas are in balance due to the discharge of the groundwater. Though wetlands also come with various benefits of ensuring that the soil is hydric and at the same time maintaining the ecosystems habitant in these areas, the increase of groundwater discharge to springs and streams dampens the amplitude of flow fluctuations thus sustaining associated wetland environments. As a control mechanism, artificial wetlands are best ways to mitigate stormwater runoff issues.

Conclusions

This chapter recommends the smart infrastructure management methods and mitigation of rainwater and runoff through practices that direct water to catchment areas with a purpose of meeting the needs of saving water and utilization of water especially for African Countries. The focus on African countries is due to some falling under dry regions which means the possibilities of these countries being faced with reduction of water in the 50 years from now would be high in case they do not observe the different methods of mitigating water. Besides, the use of these mitigation processes has socioeconomic impacts, which include the reduction of negative impacts of disaster risk management such as floods and drought on social life and reduction of energy costs, which is economical to a country.

Most countries in Africa are currently adapting the advanced technologies such as Internet of Things (IoT), and Artificial Intelligence, either, in their cities, homes, or work environments. Implementation of MAR with such technologies has a likelihood of ensuring quality, and sufficient water for most African countries, thus improving the country's economic status. The main agenda is to ensure that we avoid urban flooding in African Countries, with the implementation of infrastructures that help in redirecting runoff water to refill aquifers when already treated. This chapter suggested a number of solutions such as permeable pavements, implementation of green roofs, retention systems, rainwater harvesting methods, and use of catchment areas.

Policy-making guidelines is another way to direct constructors on how to implement certain measures when constructing a structure that could distract water flow to reach the aquifers when treated and safe. Use of these guidelines ensures to incorporate the decision making and avoid constraints such as environmental constraints (instability) and financial constraints (more expenses in water treatments and retention systems).

Water harvesting and other techniques have been used to help preserve water before runoff, the use of man-made aquifer recharge methods as a form of SUDs is a natural and more efficient way of helping runoff water to seep back to the water table and ensure a balance. This allows the table to follow its natural way of seeping in between permeable rocks from saturated to unsaturated areas. While trying to curb global warming, this is a step toward managing the effects of the melting ice caps through the natural replenishing of the water table. This practice can easily be adopted by everyone as we build more and more buildings creating a balance between the need for human settlement and the natural way of water replenishing itself.

References

Alley W, Reilly TE, Franke OL (1999) Sustainability of ground-water resources. CIR. U.S.Dept of the Interior. U.S Geological Survey. https://doi.org/10.3133/cir1186

Bache DH, Gregory R (2007) Flocs in water treatment. The International Water Association Publishing

Black J, Maimbo M, Oduor A, Cherogony K, Meerman K (2012) Rainwater harvesting inventory of Kenya. An overview of techniques, sustainability factor and stakeholders. https://doi.org/10.13140/2.1.2299.5208

Chow D (2013) Water woes: vast US aquifer is being tapped out. *Live Science*. Retrieved from https://www.livescience.com/39186-kansas-aquifer-water-depletion.html

Critical Concrete (2018, December 18) Retrieved from critical concrete website: https://criticalconcrete.com/green-roof/

Dewandel B, Ganolfi JM, De Condappa D, Ahmad S (2008) An efficient methodology for estimating irrigation return flow coefficients of irrigated crops at watershed and seasonal scale. Web of Science Core Collection 22(11):1700–1712. https://doi.org/10.1002/hyp.6738

Dillo P, Enrique F, Albert T (2011) Management of aquifer recharge and discharge processes and aquifer storage Equilibrium. Thematic Paper

Eschool (2018) Wastewater treatment types. Retrieved from https://eschool.iaspaper.net/wastewater-treatment-facts/wastewater-treatment-types/

Fluence News Team (2017, May 8) Fluence. Retrieved from Fluence: https://www.fluencecorp.com/what-is-stormwater/

Geiger WF (2020) Sustainable water management chance for mega cities in developing countries. In: International symposium on new directions in urban water management

Hidrologia Sostenible (n.d.) Hidrologia sostenible. Retrieved from Hidrologia Sostenible: http://hidrologiasostenible.com/sustainable-urban-drainage-systems-suds/

Hydrolink Technologies (2014) Hydrolink technologies. Retrieved from Hydrolink Technologies: http://hydrolink.co.ke/reverse-osmosis-systems

Kenya Data Portal (2019) Kenya data portal. Retrieved from Kenya Data Portal: https://kenya.opendataforafrica.org/

Matos JS (2003) Aspectos Históricos a Actuais da Evolução da Drenagem de Águas Residuais em Meio Urbano. Revista Engenharia Civil, Lisboa

Mays LW (2001) Stormwater collection systems design handbook. McGraw Hill Professional, New York

National parks Association (n.d.) National parks association of NSW. Retrieved from National Parks Association: http://www.bushwalking101.org/recognising-topographic-features/

National Research Council (2002) Riparian areas: functions and strategies for management. Consensus Study Report. https://doi.org/10.17226/10327

Ogwu M (2019) Towards sustainable development in Africa: the challenge of urbanization and climate change adaptation. (P. a. Cobbinah, Ed.) The geography of climate change adaptation in Urban Africa. pp. 29–55. Retrieved from https://doi.org/10.1007/978-3-030-04873-0_2

Poleto C, Tassi R (2012) Sustainable urban drainage systems. Federal University of Technology, Brazil

Reed B (2004) Sustainable urban Drainage in Low-income Countries. Water, Engineering and Development Centre. Loughborough University Leicestershire. Loughborough University, UK

Taylor RG, Koussis AD, Tindimugaya C (2009) Groundwater and climate in Africa. Hydrological Science Journal 655–664. https://doi.org/10.1623/hysj.54.4.655

The Ocean Portal Team (2018) Ocean portal. Retrieved from Ocean Portal: https://ocean.si.edu/through-time/ancient-seas/sea-level-rise

The Renewable Energy Hub UK (2018) The renewable energy hub UK. Retrieved from The Renewable Energy Hub UK Website: https://www.renewableenergyhub.co.uk/main/rainwater-harvesting-information/types-of-rainwater-harvesting/

The United Nations-Water (2015) Water for a sustainble world. United Nations Educational, Scientific and Cultural, de Fontenoy

Tim II (2008) Trabalho de Integralização Multidisciplinar II. Termo de Referência

UN Environment Programme (2020) UN environment programme. Retrieved from UN Environment Programme: https://www.unenvironment.org/regions/africa/regional-initiatives/responding-climate-change

UN Water (2013) UN - water. Retrieved from UN - Water: https://www.unwater.org/publications/water-security-infographic/

Viala E (2016) USAID. Retrieved from USAID

Willem VC (2005) Rooftop gardening

Woods BB, Wilson S, Udale-Clarke H, Illiman S, Scott T, Ashley R, Kellagher (2015) The SUDS manual. Construction Industry Research and Information Association, London

World Health Organization (2003) pH in drinking-water. Guidelines for drinking-water quality, 2. Marketing and Dissemination, World Health Organization. Geneva 27, Switzerland

12

Thermodynamic Environment During the 2009 Burkina Faso and 2012 Nigeria Flood Disasters

R. Ayodeji Balogun, E. Adesanya Adefisan, Z. Debo Adeyewa and E. Chilekwu Okogbue

Contents

Abstract

Critical or extreme atmospheric conditions which could result in flood disasters are important output for numerical weather forecast. This research applied thermodynamic variables to investigate the environment of two flood scenarios in West Africa as captured by the Tropical Rainfall Measurement Mission (TRMM) satellite. Results from the two case studies of flood events, in (i) Burkina Faso and (ii) Nigeria savannah, investigated in this research work, indicated that the September 1st 2009 flood, which was as a result of a single volumetric rainfall event of 408,070.60 ((mm/h)*km^2) with 65% convective region in Burkina Faso, was initiated by interactions between extremely large lower tropospheric wind shear and cold pool dynamics. The case of the Nigeria savannah floods between July and September, 2012, was triggered by both continuous rainfall and release

R. A. Balogun (✉) · E. A. Adefisan · Z. D. Adeyewa · E. C. Okogbue
Department of Meteorology and Climate Science, Federal University of Technology, Akure, Nigeria
e-mail: rabalogun@futa.edu.ng; eaadefisan@futa.edu.ng; adeyewa@run.edu.ng; emokogbue@gmail.com

of water from the lagdo dam in Cameroon, which affected most of the communities in the river Benue axis. The continuous rainfalls were found to be as a result of extremely high convergence of moisture in the river Benue axis at different locations and periods. One of such rainfall events, as captured by TRMM satellite during September 29, 2012 in the Nigeria rainforest zone, indicated that the volumetric rainfall is 351,310.9 ((mm/h)*km^2) with only 34% convective portion. From these results, it can be deduced that a combination of thermodynamic environmental variables, volume rainfall, and other satellite-derived convective parameters could provide important information for flood forecasting.

Keywords

Flood disaster · Thermodynamic · Environment · Volumetric rainfall · Moisture convergence

Introduction

Annually, between March and early October, severe precipitation event which are often associated with flash floods events frequently cause severe damage and loss of property and lives in most countries of West Africa (Balogun et al. 2018). This climate change hazard has been linked to pressures from socioeconomic and physical factors (Merem et al. 2019). The magnitude and frequency of climate hazards have generated serious concerns in recent years in Nigeria, since the 2012 flood disaster in the country and most other West African countries in particular. Flood disasters have been a major occurrence that keeps recurring annually, in some parts of Nigeria (Magami et al. 2014). A Post-Disaster Needs Assessment (PDNA), of the 2012 Nigeria flood, was conducted by the Federal Republic of Nigeria (FRN) with technical support from United Nations, European Union, and World Bank. The major aim of the assessment was to reduce the impact of future disasters, since the floods are due to climate change (Nigeria-PDNA 2013).

Burkina Faso, which is surrounded by other West African countries particularly in the south, shared both Savannah and Sahel climates of West Africa. The economy is specifically based on cotton production and gold mining (WB 2017). Most parts of the country, and Ouagadougou which is the capital in particular, were ravaged by a devastating flood event on the 1st of September 2009 (GBF et al. 2010). This was a result of an extreme single-day rainfall amount of 261.3 mm (Karambiri 2009; GBF et al. 2010; Galvin 2010). GBF et al. 2010 stated that the September 1st 2009 flood event affected 11 out of the 13 regions in Burkina Faso with the majority of the impacts observed to be concentrated in Ouagadougou. According to GBF et al. (2010), the flood event caused at least 46 deaths and about 120,000 victims were affected. The economic impact was estimated at $201 Million USD in damages, $33 Million USD in losses, and $266 Million USD in reconstruction needs (GBF et al. 2010). (Galvin 2010) concluded that the flood event was "*heavy,*" "*extreme,*" "*enormous,*" and "*exceptional*"; while GBF et al. (2010) concluded that the single-day rainfall represented almost one-third of the total rainfall received annually during the monsoon season. The flood event has a return period of more than 10,000 years according to a preliminary statistical analysis by Karambiri (2009).

The 2012 flood disaster in Nigeria was caused by continuous heavy rainfall and the release of water from Lagdo Dam and Lake Nyos in the Cameroon. The water released affected border towns in Nigeria (Ojobor 2014). NIMET (2012) had earlier predicted, in early March during the 2012 rainfall prediction, and cautioned the Nigerian government about the possibility of continuous rainfall, which could result in destructive flooding in most states in Nigeria, with emphasis on the river basin or catchment and coastal communities of Nigeria. NIMET (2012) advised the government of Nigeria to organize public enlightment programs so as to minimize the damages, devastation, and casualties, in terms of lives and properties, most importantly in locations and communities close to rivers, dams, among others. According to Ojobor (2014) and Agada (2015), a huge number of the populations were largely displaced in these communities, due to the 2012 flood event, as predicted by NIMET (2012). The intention was to help Government agencies and the populace to put necessary measures in place to mitigate the effects of flood waters (Ojobor 2014). The second warning came in August 2012 from Cameroun. Akaeze (2012) stated that the Nigerian Government was officially written by the Cameroon government that the Lagdo Dam located in Cameroun has already filled to capacity as a result of continuous rainfall in the location of the dam and if such rainfall amount should continue in the following week, the dam will overflow and will result in recurrent overflow of water (Akaeze 2012). According to Njoku (2013), the Cameroon authority referenced paragraph 3 of article 111, in the 2010 Nigeria–Cameroon Joint Commission regarding the interchange of relevant information as regard the situations around the Lake Nyos and Lagdo Dam, so that the Nigerian Government will take precautionary measures around the flood areas so as to avoid severe loss.

This chapter was contributed in 2019 to investigate the atmospheric environment of the two case studies discussed above. The major limitation was lack of ground observations to validate the satellite data used for this chapter.

The TRMM and ERA: Interim Data Description

This chapter investigates high resolution (0.1 × 0.1) TRMM 2A25 radar-derived rainfall product, alongside derived products of the European Centre for Medium-range Weather Forecasts (ECMWF) interim reanalysis (ERA-Interim: Simmons et al. 2007; Dee et al. 2011) dataset to investigate the two (2) case studies.

The TRMM orbital data provided date, time, location, and orbit number analyzed in the case studies. The parameters analyzed from the TRMM orbital data are: Infrared (10.8 micron) brightness temperature, near-surface reflectivity, 85 GHz PCT (Polarization Corrected Temperature), and near-surface rain rate. The parameters analyzed from the environment, of ERA-Interim reanalysis dataset are Sounding (skew – T) to further provide information on convective development, convective available potential energy (CAPE), moisture fluxes, lifted index, and vertical wind shear. The environmental parameters were analyzed at 6-hour intervals a day before, during, and a day after the case study event(s).

The 2009 Flood in Burkina Faso

The severity of the September 1st 2009 flood made it a focal point of this chapter. The TRMM snapshot of a Mesoscale Convective System (MCS) on September 1, 2009, covering southern and western Burkina Faso, is shown in Fig. 1a–d. The Visible Infrared Scanner (VIRS), in the TRMM satellite, estimated the brightness temperature as low as 178 K in Fig. 1a. The pattern of the MCS, as shown by the near surface reflectivity (Fig. 1b), 85 GHz Polarization Corrected Temperature (PCT, Fig. 1c), and the near surface rainfall (Fig. 1d), clearly indicated that the system is a line squall which spans several kilometers. The rain rate from the single system is over 130 mm/h and the MCS is only 65% convective (Table 1), meaning that the trailing stratiform portion (region) within the system is 35%. The volume of rainfall (volumetric rainfall) from the single system is 408,070.60 ((mm/h)*km^2) (Table 1). The total number of flashcount from the single system is 191 as estimated by the Lightning

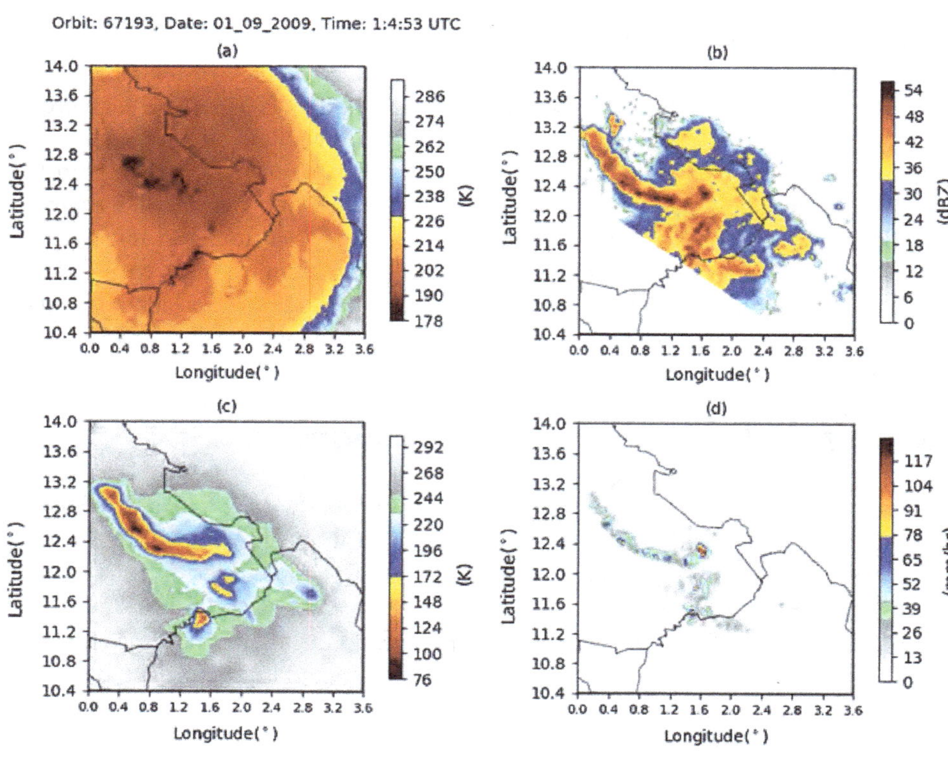

Fig. 1 Case study of TRMM-observed MCS which occurred on 1st September 2009 in Burkina Faso noted **(a)** Infrared (10.8 micron) brightness temperature (Tb), **(b)** PR near surface reflectivity, **(c)** 85GHz PCT of MCS, and **(d)** 2A25 near surface rain rate

Table 1 Summary of some parameters from the observed system during September 1, 2009

Latitude	Longitude	% Convective	Size (Km^2)	Volrain ((mm/h) *km^2)	Flashcount (#)
12.16	1.64	65	52116.35	408070.6	191

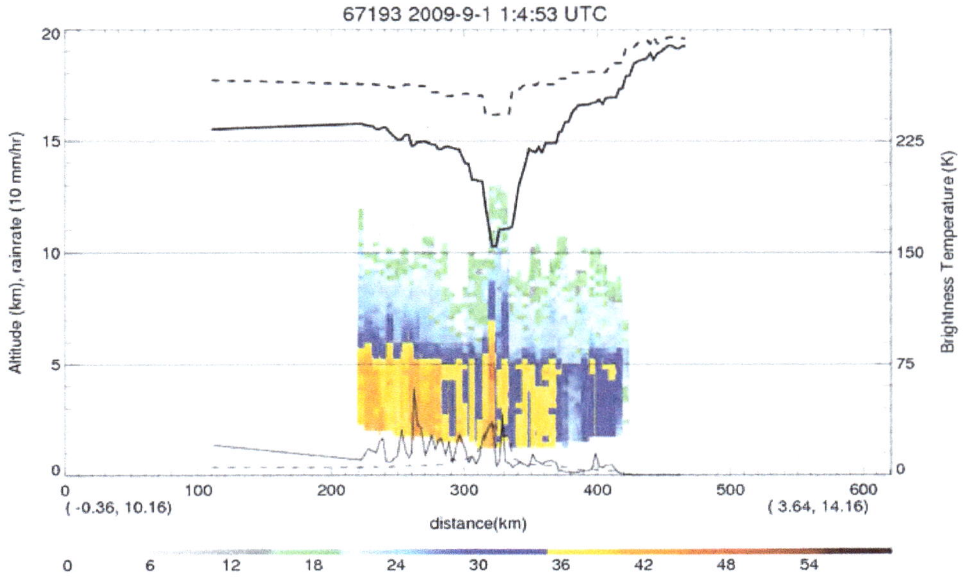

Fig. 2 Vertical cross section of the MCS which occurred on September 1, 2009

Imaging Sensor (LIS) of the TRMM satellite. The system was captured at exactly 1:04:53 am on the 1st of September 2009 by the TRMM satellite. The vertical cross section, in Fig. 2, shows that the system is about 200 km horizontally and reaches about 13–14 km altitude, which is close to the tropopause.

The thermodynamic parameters considered in this chapter to discuss the environment of this system are moisture flux convergence at the 1000 mb (hpa) level, lifted index, wind shear between 1 and 3 km, and the skewt diagram (which will provide information on the Lifting Condensation Level (LCL), Level of Free Convection (LFC), Cloud Base (CB), Cloud Top (CT), among others). The environment of the system was investigated a day before the system was captured, the day the system was captured, and the day after the system was captured.

On the August 31, 2009, a day before the flood, the location where this system occurred clearly indicated weak moisture divergence (tendency of a clear sky), in the range of 0–10 kg m^{-2} s^{-1} occurring in the zone, but south of this zone is the presence of strong moisture convergence (mechanism by which air is lifted for cloud formation) in the range of 55–80 kg m^{-2} s^{-1}. A close observation of the time periods indicates that the moisture convergence seem to be migrating northward toward the location of the system. At 18Z, part of the location of the system is already experiencing weak moisture convergence. During the night time of September 1,

2009, there was a stronger moisture convergence south of the flood event zone, but the moisture convergence was still weak in the flood event zone. Other periods (6Z, 12Z, and 18Z) did not show strong moisture convergence on the 1st of September 2009, which was the day of the flood. On 2nd of September 2009, the entire environment in which the flood occurred encountered moisture divergence, which indicated that the environment is already dry or in clear sky condition. From the analysis of the moisture flux convergence in the location of the flood, there is tendency that the system developed south of the location, and migrated northward into the zone.

The lifted index (LI), similar to the Convective Available Potential Energy (CAPE), is used to assess low level parcel (in)stability of the troposphere. A negative lifted index (LI) indicates that the boundary layer is unstable with respect to the middle troposphere. Such is an environment in which convection can occur. The more negative the LI, the more unstable the troposphere and the more buoyant the acceleration will be for rising parcels of air from the Planetary Boundary Layer (PBL). On the 31st of August 2009, the environment experienced weak instability during the night, morning, and day time, whereas during the evening time, the environment became more unstable. On the 1st of September 2009, the night time, which was preceded by the evening period of 31st August 2009, showed that the environment of the flood event location was more stable (lesser instability) than the preceding evening. A day after the flood scenario, the stability of the environment was similar to the previous day. From these discussions so far on (in)stability in the environment of the flood scenario using lifted index, it was obvious that the environment was not strongly unstable during the flood scenario event of 1st September 2009. The only case where large instability was observed, in the range of −4 to −5 K, was during the evening period which preceded the night time when the flood scenario occurred. In all the other periods, only marginal instabilities were observed, which may not be strong enough for buoyant acceleration of air parcels in the Planetary Boundary Layer (PBL).

The question from the foregoing is can deep convective clouds develop under this kind of environment in which the moisture convergence and instabilities were relatively weak? The zone where the flood event occurred was in the Sudano and it occurred during the JJAS season, which is the season for strong convective storms in the zone.

Vertical wind shear is an important factor for determining the development of Mesoscale Convective Systems (MCSs, Chen et al. 2015). Many studies have examined the effects of vertical wind shear on intensity and organization of convective systems, especially for squall lines (Rotunno et al. 1988; Weisman and Rotunno 2004; Coniglio et al. 2006; Takemi 2006), similar to the event being examined in this section. The effects of strong wind shear on cloud system organization depend on the altitudes where the strong wind shear occurs. Lower level shear is also favorable for organizing systems into lines or quasi-lines (Wu and Yanai 1994; Robe and Emanuel 2001; Richardson et al. 2007; DeLonge et al. 2010), as a reminder that the TRMM snapshot earlier presented showed this particular system to be an organized MCS (squall line). The vertical low level (1–3 km) wind shear seemed to gradually developed into a severe (>8 s^{-}) scale wind shear, from the night (0Z) of August 31, 2009 through to the evening period (18Z) of 31st August 2009, in the location of the flood event. This strong (severe) low level vertical wind shear, which is most severe in the eastern axis of the flood location, extends into the flood location during the evening period of 31st August 2009 and night time (0Z) of 1st September 2009 (which was the exact time the MCS was captured). This severe low level vertical wind shear could have initiated the development and evolution of convective cloud, and hence MCS similar to the finding that lower level shear directly influences the initiation and evolution of convective clouds due to interactions between lower tropospheric shear and cold pool dynamics (Rotunno et al. 1988; Moncrieff 1992; Xu et al. 1992), which could have originated from the 1000 mb moisture convergence located southward and migrated northward into the location of the flood event. Historical studies (Simpson 1969; Charba 1974) have shown that a region/zone dominated by evaporatively cooled air which was transported to lower levels via convective downdrafts and thereafter, upon reaching the surface, spread out in the form of a density current is referred to as a convective cold pool region. Goff (1976) stated that cold pools often displaced warmer environmental air as they spread out to adjacent zones. This displaced ambient air is often buoyant and are then uplifted, and thus results in the development or initiation of new convective clouds (Goff 1976; Warner et al. 1980). Droegemeier and Wilhelmson (1985) had noted that two or more collisions of cold pools are effective trigger mechanism for initiation of convection. In addition, Tompkins (2001) and Langhans and Romps (2015) stated in their works that cold pools often bring surface gusty winds that intensify sensible and latent heat fluxes, and thus modifying the thermodynamic characteristics and moisture structure of the boundary layer. The vertical wind shear subsides slightly, although still in the severe category, but maintained the severe category up till the night time (0Z) of the next day (September 2, 2009). Thereafter, the low level wind shear dropped to moderate category on the morning (6Z) of September 2, 2009 and other periods (12Z and 18Z) of the same day.

By the interaction between low level wind shear and cool pools in convective cloud initiation and development (or evolution), especially in the roles these parameters play in the Burkina Faso flood event of 1st September 2009, the next assessment of the case study is to investigate the cloud structure through the use of skew-t diagram.

Results showed that the Convective Available Potential Energy (CAPE) environment was positive (+) for all the periods (0Z, 6Z, 12Z, and 18Z) on the 31st August 2009, with the peak mean CAPE value (1392.42 J/kg) indicated by the evening period (18Z), which was the period that preceded the night time (0Z) of 1st September 2009 when the MCS system was captured. During the night time (0Z) when the MCS system was captured, the top of the cloud is around 13–14 km similar to the height estimated by the TRMM satellite in the cross section plot discussed earlier. The mean CAPE value dropped significantly during the morning (6Z), day (12Z), and evening (18Z) periods of the same day and through all the periods (0Z, 6Z, 12Z, and 18Z) of the second day, which was September 2, 2009.

Records from the flood, which lasted in the memory of the inhabitant, has it that 180, 386 victims were affected, 62 victims were wound, 41 died, and 33,172 homes were destroyed (Schlef 2018).

The 2009 flood created a lasting and strongly negative impression on the victims; they described the flood as "*catastrophic*" and "*big*" and one public official called floods "*bitter experiences*" while a flood victim said "*what we lived through, we have no desire to experience again*" (Schlef 2018). The flood victims and some public officials also shared their personal stories, revealing the unquantifiable and sometimes long-lasting effects of experiencing a devastating flood: "*We left the house ... it overflowed ... I didn't know how to swim, so they dragged me along ... even now when it rains at night I can't sleep – I'm afraid! And I'm not the only one*" (Schlef 2018).

The 2012 Flood in Nigeria

Between July and October 2012, floods which resulted from the predicted heavy rains and the released water from Lagdo dam in Cameroon pushed rivers over their banks and submerged hundreds of thousands of acres of farmland. By mid-October, these floods had forced 1.3 million people from their homes and claimed 431 lives (Agada 2015). The severity of the 2012 Nigeria flood made it a focal point of this chapter. The TRMM snapshot of a Mesoscale Convective System (MCS), with an extreme volume rainfall (Table 2) of 351,310.9 ((mm/h)*km^2) on 29th September 2012 (one of the days of the flood), in Ofutop II (a community close to Ikom Cross River state) is shown in Fig. 3. The Visible Infrared Scanner (VIRS), in the TRMM satellite, estimated the brightness temperature as low as 186 K in Fig. 3a. The rain rate from the single system is over 140 mm/h and the MCS is only 34.1% convective (Table 2), meaning that the trailing stratiform portion (region) within the system is 65.9%. The total number of flash count from the single system is 5 (Table 2) as estimated by the Lightning Imaging Sensor (LIS) of the TRMM satellite. The system was captured at exactly 7:38:20 am on the 29th of September 2012 by the TRMM satellite. By the last three (3) sentences (stratiform region was 65.9%, only 5 flashcount was estimated, and the system occurred in the morning diurnal), clearly this system is more of stratiform than convective, which would have lasted for a long time before it completely decayed. The vertical cross section in Fig. 4 shows that the system extends over 200 km horizontally and reached only about 8 km altitude in the atmosphere, and no depressed brightness temperature was observed in the cross section of the system.

As was analyzed for the Burkina Faso flood scenario, the thermodynamic parameters to investigate the environment of this system are: moisture flux convergence at the 1000 mb (hpa) level, lifted index, wind shear between 1 and 3 km, and the skewt diagram (which will provide information on the Lifting Condensation Level (LCL),

Table 2 Summary of some parameters from the observed system during 29th September 2012

Latitude	Longitude	% convective	Size (km^2)	Volrain ((mm/h)*km^2)	Flashcount (#)
6.78	7.13	34.1	83190.8	351310.9	5

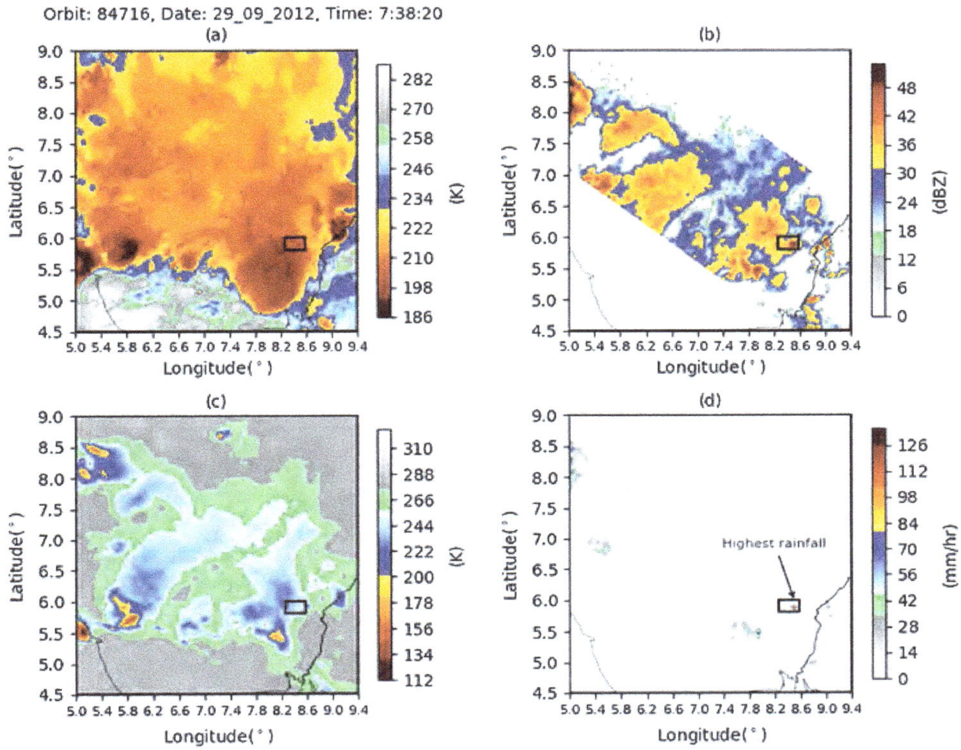

Fig. 3 Case study of TRMM-observed MCS which occurred on September 29, 2012 in Nigeria, where (**a**) Infrared (10.8 micron) brightness temperature (Tb), (**b**) PR near surface reflectivity, (**c**) 85GHz PCT of MCS, and (**d**) 2A25 near surface rain rate

Level of Free Convection (LFC), Cloud Base (CB), Cloud Top (CT), among others). For this case study, the environment of the system was investigated a day before the system was captured, the day the system was captured, and five (5) additional days after the system was captured because the flood affected many parts of Nigeria, and was reported to last many days beginning from around July to around mid-October.

On 28th of September 2012, a day before the system was captured by the TRMM satellite, extreme low level moisture flux convergence was observed during all the periods (0Z, 6Z, 12Z and 18Z), especially along the axis of the river Benue bank, which were the most affected areas (the Nigeria savannah zone). The communities in Nigeria affected by this extreme low level moisture flux convergence are Aguleri in Anambra state, Katsina Ala in Benue state, Makurdi in Benue state, Ikom in Cross River state, Lokoja in Kogi state, among others. It has been noted by Ismail and Siadari (2017) that if the atmosphere is unstable with abundant low-level moisture and a mechanism exists to lift the air (thereby releasing the potential instability), convective weather and rainfall (showers) can develop. On September 29, 2012, the affected areas were still under the influenced of extreme low-level moisture flux convergence, including Ofutop II in Cross River state where the peak rainfall rate was observed. On September 30, 2012, Aguleri community in Anambra state was

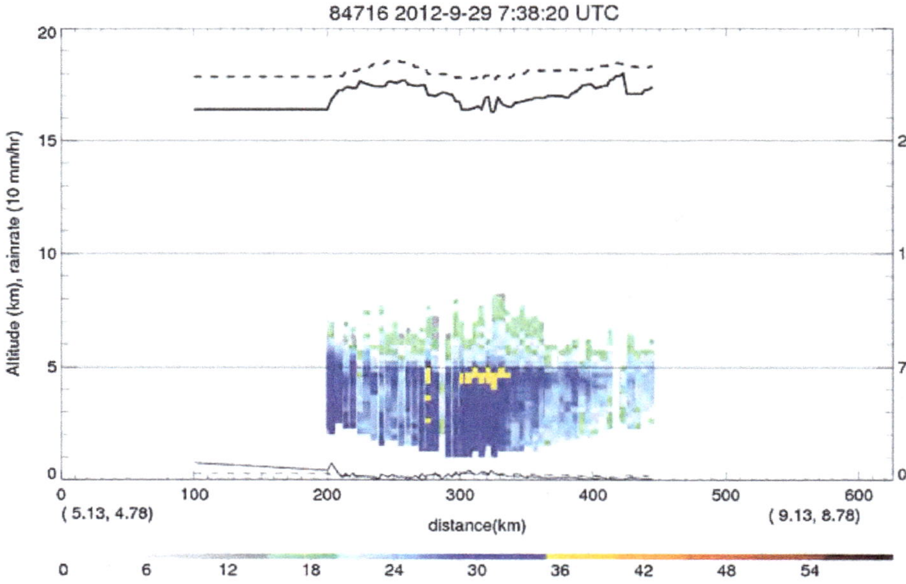

Fig. 4 Vertical cross section of the MCS which occurred on September 29, 2012

under the influence of extreme low level moisture flux convergence during the night time. According to Vanguard newspaper of 7th October 2012, the then governor of Anambra state, Governor Peter Obi, had to suspend the Independence Day celebration of Nigeria, slated for 1st of October 2012, so as to pass through the flooded Onitsha-Osomala road in Nmiata and Ogbaru road in Anambra West in other to supervise the removal of victims trapped in their communities by the devastating flood event. Vanguard newspaper of 7th October 2012 further stated that the entire Nzam, Aguleri Otu (a community where the then President of Nigeria, Goodluck Jonathan, commissioned the Orient Petroleum facility), Nmiata, Odekpe, Osomala, Osuche, Atani, Amii, Umuzu, and other communities in Anambra state were under water. Properties that worth billions of naira in the four local government areas of Anambra state, where these communities were situated, were destroyed by the flood. According to the Vanguard of 7th October 2012, the governor was able to persuade the people to leave their properties and secure their lives. He, the then Governor Peter Obi, thereafter used canoe to visit other communities trapped across the River Niger basin of Anambra state. By the time the then governor Peter Obi returned to his entourage to continue his inspection of other affected communities, flood had already filled the whole road through which he would leave. The governor and his entourage had to walk through the flood (Fig. 5a), which leveled-up to their waists, in other to reach the convoy of vehicles. On 1st through to 4th October 2012, all the communities along the River Benue axis were still under the influence of extreme low level moisture flux convergence. This implied that the communities would have experienced continuous heavy precipitation during these periods, apart from the water that was released from the lagdo dam (see Fig. 6) in Cameroon.

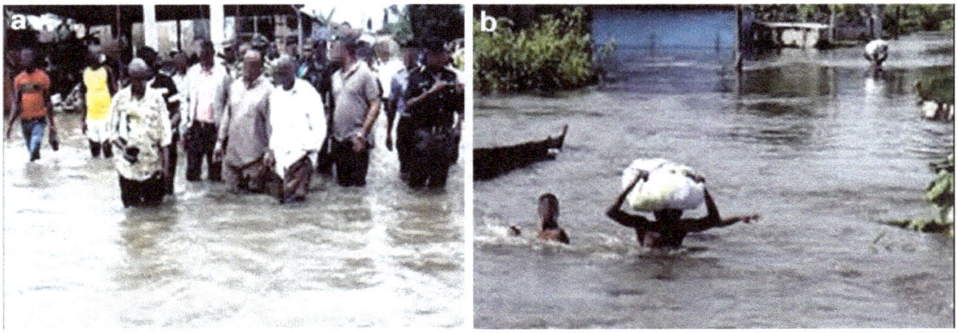

Fig. 5 Situations during the 2012 flood in Anambra state (**a** and **b**). (Source: Emmanuel et al. 2015)

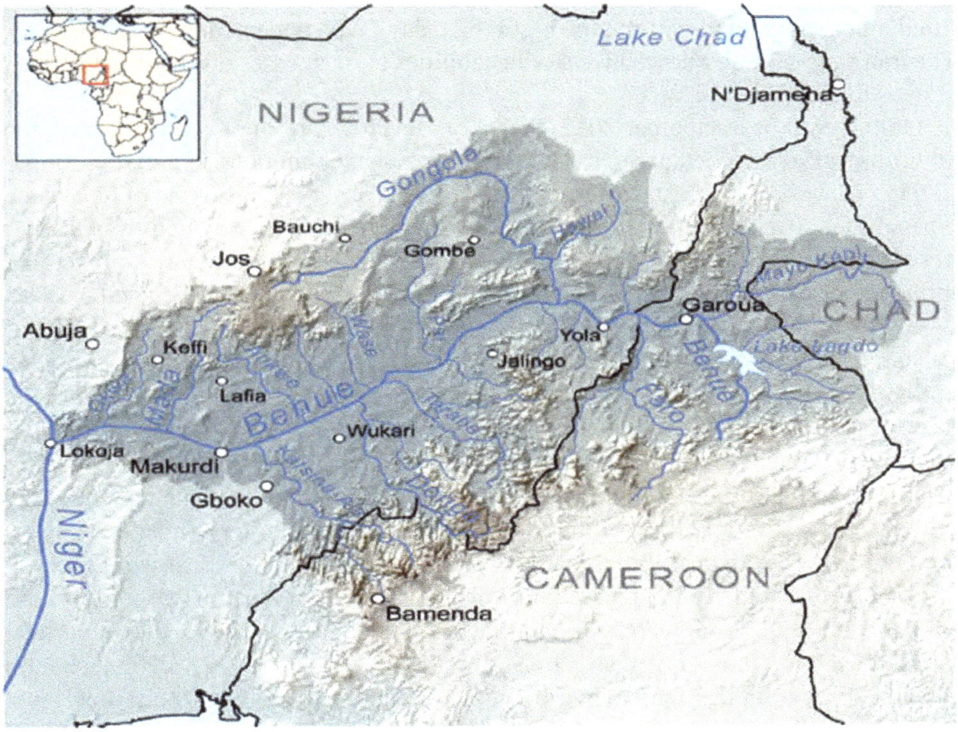

Fig. 6 Map showing Nigeria, Cameroon, Benue River, and Lake Lagdo in Cameroon.

It was also reported (Vanguard, October 7, 2012), that about 79, 000 people were displaced in Cross River State. Out of the number, more than half – 49, 918 – were displaced by flood as a result of the release of water from Lagdo Dam in Cameroon. In Ikom Local Government Area, close to the community where the peak rainfall was observed on 29th September 2012, the flood also displaced 26 communities and destroyed one school and two markets.

The stability of the environment, on 28th September 2012 indicated that the entire zone under investigation was extremely unstable, particularly during the evening time (18Z) of the day that preceded the MCS event that was captured by TRMM satellite. It clearly showed that the Ofutop II community in Cross River was extremely unstable during the evening period of 28th September 2012 and during the night (0Z) of September 29, 2012 (the day the MCS was captured by TRMM satellite). During the morning time (6Z), the instability of the environment in Ofutop II of Cross River state had subsided to moderate category, but increased again to severe category during the day (12Z) and evening (18Z) periods. On September 30, 2012, the day that preceded the governor's visit to communities, Aguleri in Anambra state was under extreme instability during the evening period (18Z). Places like Makurdi and Katsina Ala in Benue state were also under extreme instability during the same evening diurnal. On 1st through to 4th October 2012, most communities in the affected zone were under moderate to extreme instability as indicated by the lifted index, particularly during the night (0Z), day (12Z) and evening (18Z) periods. The morning diurnals showed weaker instabilities in most cases during these periods (1st – 4th October, 2012).

On the 28th of September 2012, there was severe – extreme – wind shear in the environment of the event, which cut across the seven communities under investigation in this work. The most extreme wind shear was observed in Lokoja of Kogi state during the evening period (18Z) of 28th September 2012, which was earlier observed during the night period (6Z) of the same day. Lokoja was under the influence of extreme low level vertical wind the second day, 29th September 2012 during the night (0Z) and evening (18Z) periods. Because the entire Nigeria savannah environment was under the influence of strong moisture convergence during this period, hence the presence of cool pool dynamics in combination with the extreme low level vertical wind shear, among other factors would definitely had initiated deep convective systems, particularly in Lokoja and surrounding Benue state. Ofutop II and surrounding Ikom, in Cross River state, were particularly under the influence of severe low level vertical wind shear on the 28th September 2012 in all the periods. On 29th and 30th September 2012, the low level vertical wind shear in Ofutop II was already suppressed. It was noted that Lokoja in Kogi state and Makurdi in Benue state were still under the influence of extreme low level vertical wind shear on 30th September 2012. On 1st of October 2012, all the Nigeria savannah communities under investigation were influenced by extreme low level vertical wind shear. This convective environment remained under the influence of extreme low level vertical wind shear during 2nd, 3rd, and 4th of October 2012. Agada (2015) stated that the two most affected areas were Kogi and Benue States.

Results for the Ofutop II community where the peak rainfall rate was captured by the TRMM satellite showed that the Convective Available Potential Energy (CAPE) environment was positive (+) for the night time (0Z) on the 28th September 2012, with the peak mean CAPE value (4672.80 J/kg) and highest cloud top (slightly above 15 km) indicated by the evening time (18Z). During the night time (0Z) of 29th September 2012, which was the diurnal that preceded the morning time (0Z) of 29th September 2012 when the MCS system was captured, the top of the cloud as

Fig. 7 Areal view of Lokoja, Kogi State, under the siege of flood. (Source: Vanguard 2012)

shown by the skew-t diagram was around 14–15 km, quite different from the height estimated by the TRMM satellite in the cross section plot, because the location (Ofutop II in Cross River state) under investigation is different from the nadir latitude–longitude of observation by the TRMM satellite. The mean CAPE value dropped significantly during the morning (6Z), day (12Z), and evening (18Z) periods of the same day but increased during the day (12Z) and evening (18Z) periods (12Z and 18Z) of the second day, which was 30th September 2012.

According to Vanguard (2012), horror came in the shape of suicide in one of the resettlement camps in Kogi State where two victims of the ravaging floods decided to take their own lives. They could not bear the loss of properties (Vanguard 2012). In relief camps in Kogi, 20 people have already died. The conditions were worst, two bags of rice for 2000 victims per day: How would it go round? The flood ravaged Lokoja (Vanguard 2012) as shown in Fig. 7 above.

Conclusions

A combination of TRMM observed convective properties and reanalysis dataset have provided valuable information for analyzing flood scenarios in West Africa and have proved reliable based on the results generated from this research work.

At the time of written this chapter, flood disasters has become a yearly occurrence in all the locations affected by the 2012 Nigerian flood. There is no specific climate change adaptation strategies adopted in any of these locations. The inhabitants watched their properties and farmlands destroyed every year. There cry had always been to the government, but in all, relief materials are what they get in return.

Because this climate change hazard has become an annual occurrence in some parts of Nigeria and West Africa in general, this work therefore suggest that governments in affected countries put in place mitigation strategies to reduce the impact of future occurrence.

Acknowledgements This research was supported by Tertiary Education Trust Fund (TETFund) grant of the Nigerian government under the academic staff development program of the Federal University of Technology Akure for a split-site research program with the Texas A & M University Corpus Christi United States. Special thanks to Dr. Chuntao Liu, who granted me access to the Precipitation Feature database at Texas A & M University Corpus Christi United States.

References

Agada S (2015) A serious flooding event in Nigeria in 2012 with specific focus on Benue State: a brief review. Nat Hazards 77:1405–1414. https://doi.org/10.1007/s11069-015-1639-4

Akaeze A (2012) An avoidable disaster. Tell special report, October 8, 2012.

Balogun RA, Liu C, Debo AZ, Okogbue EC, Adesanya AE (2018) Intraseasonal and seasonal variability of convective properties of monsoon precipitation systems over West and Central Africa. Theor Appl Climatol. https://doi.org/10.1007/s00704-018-2692-1

Charba J (1974) Application of gravity current model to analysis of squall-line gust front. Mon Weather Rev 102(2):140–156

Chen Q, Fan J, Hagos S, Gustafson WI Jr, Berg LK (2015) Roles of wind shear at different vertical levels: cloud system organization and properties. J Geophys Res Atmos 120:6551–6574. https://doi.org/10.1002/2015JD023253

Coniglio MC, Stensrud DJ, Wicker LJ (2006) Effects of upper-level shear on the structure and maintenance of strong quasi-linear mesoscale convective systems. J Atmos Sci 63(4):1231–1252. https://doi.org/10.1175/jas3681.1

Dee DP et al (2011) The ERA-Interim reanalysis: configuration and performance of the data assimilation system. Q J Roy Meteor Soc 137:553–597

DeLonge MS, Fuentes JD, Chan S, Kucera PA, Joseph E, Gaye AT, Daouda B (2010) Attributes of mesoscale convective systems at the land-ocean transition in Senegal during NASA African monsoon multidisciplinary analyses 2006. J Geophys Res 115:D10213. https://doi.org/10.1029/2009JD012518

Droegemeier KK, Wilhelmson RB (1985) Three-dimensional numerical modeling of convection produced by interacting thunderstorm outflows. Part I: control simulation and low-level moisture variations. J Atmos Sci 42(22):2381–2403. https://doi.org/10.1175/1520-0469(1985)042<2381:TDNMOC>2.0.CO;2

Emmanuel Udo A, Ojinnaka OC, Baywood CN, Gift UA (2015) Flood hazard analysis and damage assessment of 2012 flood in Anambra state using GIS and remote sensing approach. Am J Geogr Inform Syst 4(1):38–51. https://doi.org/10.5923/j.ajgis.20150401.03

Galvin JFP (2010) Two easterly waves in West Africa in summer 2009. Weather 65(8):219–227. https://doi.org/10.1002/wea.605.

GBF, WB, UN (2010) Inondations du 1er Septembre 2009 au Burkina Faso: evaluation des dommages, pertes et besoins de construction, de reconstruction et de relevement (Vol. 56803 BF). Government of Burkina Faso, World Bank, United Nations

Goff RC (1976) Vertical structure of thunderstorm outflows. Mon Weather Rev 104(11):1429–1440. https://doi.org/10.1175/1520-0493(1976)104<1429:VSOTO>2.0.CO;2

Ismail P, Siadari EL (2017) Atmospheric conditions analysis of the heavy rain phenomenon in Biak (case study 4–5 December 2014). J Phys Conf Ser 795:012015

Karambiri H (2009) Breve analyse frequentielle de la pluie du 1er septembre 2009 a Ouagadougou (Burkina Faso). Ouagadougou: 2iE.

Langhans W, Romps DM (2015) The origin of water vapor rings in tropical oceanic cold pools. Geophys Res Lett 42:7825–7834. https://doi.org/10.1002/2015GL065623

Magami I, Yahaya S, Mohammed K (2014) Causes and consequences of flooding in Nigeria: a review. Biol Environ Sci J Tropics 11(2):154–162

Merem E, Twumasi Y, Wesley J, Alsarari M, Fageir S, Crisler M, Romorno C, Olagbegi D, Hines A, Ochai G, Nwagboso E, Leggett S, Foster D, Purry V, Washington J (2019) Regional assessment of climate change hazards in Southern Nigeria with GIS. J Saf Eng 8(1):9–27

Moncrieff MW (1992) Organized convective systems: archetypal dynamical models, mass and momentumflux theory, and parametrization. Q J R Meteorol Soc 118(507):819–850. https://doi.org/10.1002/qj.49711850703

NIMET (2012) Seasonal rainfall prediction and socio-economic implications for Nigeria, Nigerian meteorological agency. http://nimet.gov.ng/sites/default/files/publications/2012-seasonal-rainfall prediction.pdf

Njoku J (2013) 2012 year of flood fury: a disaster foretold, but ignored? Vanguardngrnews, Saturday August 17, 2013.

Ojobor L (2014) The 2012 flood disaster in Nigeria: impact on the operations of small scale business women in Igboland. Int J Bus Manag Stud 3(1):463–481

Post-Disaster Needs Assessment 2012 Floods (2013) A report by the Federal Government of Nigeria with technical support from the World Bank, EU, UN, and Other Partners

Richardson YP, Droegemeier KK, Davies-Jones RP (2007) The influence of horizontal environmental variability on numerically simulated convective storms. Part I: variations in vertical shear. Mon Weather Rev 135(10):3429–3455. https://doi.org/10.1175/mwr3463.1

Robe FR, Emanuel KA (2001) The effect of vertical wind shear on radiative–convective equilibrium states. J Atmos Sci 58(11):1427–1445. https://doi.org/10.1175/1520-0469(2001)058<1427:teovws>2.0.co;2

Rotunno R, Klemp JB, Weisman ML (1988) A theory for strong, long-lived squall lines. J Atmos Sci 45(3):463–485. https://doi.org/10.1175/15200469(1988)045<0463:atfsll>2.0.co;2

Schlef K (2018) Flood risk assessment, management and perceptions in a changing world. Doctoral dissertations. 1273. https://scholarworks.umass.edu/dissertations_2/1273

Simmons A, Uppala S, Dee D, Kobayashi S (2007) ERA-interim: new ECMWF reanalysis products from 1989 onwards. ECMWF Newsl 110:25–35

Simpson JE (1969) A comparison between laboratory and atmospheric density currents. Q J R Meteorol Soc 95(406):758–765. https://doi.org/10.1002/qj.49709540609

Takemi T (2006) Impacts of moisture profile on the evolution and organization of midlatitude squall lines under various shear conditions. Atmos Res 82(1–2):37–54. https://doi.org/10.1016/j.atmosres.2005.01.007

Tompkins AM (2001) Organization of tropical convection in low vertical wind shears: the role of cold pools. J Atmos Sci 58(13):1650–1672. https://doi.org/10.1175/15200469(2001)058<1650:OOTCIL>2.0.CO;2

Vangard Newspaper (2012) Rage of nature: flood ravages communities across Nigeria. 7th October 2012

Warner C, Simpson J, Van Helvoirt G, Martin DW, Suchman D, Austin GL (1980) Deep convection on day 261 of GATE. Mon Weather Rev 108(2):169–194. https://doi.org/10.1175/1520-0493(1980)108<0169:DCODOG>2.0.CO;2

WB (2017) Burkina Faso overview. Retrieved January 1, 2017, from http://www.worldbank.org/en/country/burkinafaso/overview

Weisman ML, Rotunno R (2004) "A theory for strong long-lived squall lines" revisited. J Atmos Sci 61(4):361–382. https://doi.org/10.1175/15200469(2004)061<0361:atfsls>2.0.co;2

Wu X, Yanai M (1994) Effects of vertical wind shear on the cumulus transport of momentum: observations and parameterization. J Atmos Sci 51(12):1640–1660. https://doi.org/10.1175/1520-0469(1994)051<1640:eovwso>2.0.co;2

Xu K-M, Arakawa A, Krueger SK (1992) The macroscopic behavior of cumulus ensembles simulated by a cumulus ensemble model. J Atmos Sci 49(24):2402–2420. https://doi.org/10.1175/1520-0469(1992)049<2402:tmboce>2.0.co;2

Permissions

All chapters in this book were first published by Springer; hereby published with permission under the Creative Commons Attribution License or equivalent. Every chapter published in this book has been scrutinized by our experts. Their significance has been extensively debated. The topics covered herein carry significant findings which will fuel the growth of the discipline. They may even be implemented as practical applications or may be referred to as a beginning point for another development.

The contributors of this book come from diverse backgrounds, making this book a truly international effort. This book will bring forth new frontiers with its revolutionizing research information and detailed analysis of the nascent developments around the world.

We would like to thank all the contributing authors for lending their expertise to make the book truly unique. They have played a crucial role in the development of this book. Without their invaluable contributions this book wouldn't have been possible. They have made vital efforts to compile up to date information on the varied aspects of this subject to make this book a valuable addition to the collection of many professionals and students.

This book was conceptualized with the vision of imparting up-to-date information and advanced data in this field. To ensure the same, a matchless editorial board was set up. Every individual on the board went through rigorous rounds of assessment to prove their worth. After which they invested a large part of their time researching and compiling the most relevant data for our readers.

The editorial board has been involved in producing this book since its inception. They have spent rigorous hours researching and exploring the diverse topics which have resulted in the successful publishing of this book. They have passed on their knowledge of decades through this book. To expedite this challenging task, the publisher supported the team at every step. A small team of assistant editors was also appointed to further simplify the editing procedure and attain best results for the readers.

Apart from the editorial board, the designing team has also invested a significant amount of their time in understanding the subject and creating the most relevant covers. They scrutinized every image to scout for the most suitable representation of the subject and create an appropriate cover for the book.

The publishing team has been an ardent support to the editorial, designing and production team. Their endless efforts to recruit the best for this project, has resulted in the accomplishment of this book. They are a veteran in the field of academics and their pool of knowledge is as vast as their experience in printing. Their expertise and guidance has proved useful at every step. Their uncompromising quality standards have made this book an exceptional effort. Their encouragement from time to time has been an inspiration for everyone.

The publisher and the editorial board hope that this book will prove to be a valuable piece of knowledge for researchers, students, practitioners and scholars across the globe.

List of Contributors

Byron Zamasiya, Kefasi Nyikahadzoi and Billy Billiard Mukamuri
Centre for Applied Social Sciences, University of Zimbabwe, Harare, Zimbabwe

Sizwile Khoza, Dewald van Niekerk and Livhuwani Nemakonde
Unit for Environmental Sciences and Management, African Centre for Disaster Studies, North-West University, Potchefstroom, South Africa

Winnie Ntinyari and Joseph P. Gweyi-Onyango
Department of Agricultural Science and Technology, Kenyatta University, Nairobi, Kenya

Gandome Mayeul L. D. Quenum
West African Science Service Centre for Climate Change and Adapted Land Use (WASCAL) Graduate Research Program in West African Climate System (GRP-WACS), Federal University of Technology, Akure (FUTA), Akure, Nigeria

Nana A. B. Klutse
Ghana Space Science and Technology Institute, Atomic Energy Commission, Accra, Ghana

Eric A. Alamou, Emmanuel A. Lawin,
Laboratory of Applied Hydrology (LHA), National Institute of Water (NIW), Cotonou, Bénin

Philip G. Oguntunde
Department of Agricultural and Environmental Engineering, Federal University of Technology, Akure, Nigeria

Isaac Larbi
Climate Change andWater Resources,West African Science Service Centre on Climate Change and Adapted Land Use (WASCAL), Université d'Abomey-Calavi, Cotonou, Benin
School of Sustainable Development, University of Environment and Sustainable Development, Somanya, Ghana
Department of Civil Engineering, Faculty of Engineering, Koforidua Technical University, Koforidua, Ghana

Clement Nyamekye
Department of Civil Engineering, Faculty of Engineering, Koforidua Technical University, Koforidua, Ghana

Fabien C. C. Hountondji
Faculté d'Agronomie, University of Parakou, Parakou, Benin

Gloria C. Okafor
Department of Civil Engineering, Nigeria Maritime University, Delta-State, Nigeria

Peter Rock Ebo Odoom
Climate Change andWater Resources,West African Science Service Centre on Climate Change and Adapted Land Use (WASCAL), Université d'Abomey-Calavi, Cotonou, Benin

Sabine Troeger
Department for Development Research, Geography Institute, University of Bonn, Bonn, Germany

Alice Nyawira Karuri
School of Humanities and Social Sciences, Strathmore University, Nairobi, Kenya

D. M. G. Njarui, M. Gatheru,
Kenya Agricultural and Livestock Research Organization (KALRO), Katumani, Kenya

S. R. Ghimire
The Biosciences eastern and central Africa – International Livestock Research Institute

T. B. Hammed and M. K. C. Sridhar
Department of Environmental Health Sciences, Faculty of Public Health, College of Medicine, University of Ibadan, Ibadan, Nigeria

Faith Ngum
Lilongwe, Malawi

Johan Bastiaensen
Institute of Development Policy (IOB), University of Antwerp, Antwerp, Belgium

Getrude Gichuhi and Stephen Gitahi
Department of Research and Innovation, Strathmore University, Nairobi, Kenya

R. Ayodeji Balogun, E. Adesanya Adefisan, Z. Debo Adeyewa and E. Chilekwu Okogbue
Department of Meteorology and Climate Science, Federal University of Technology, Akure, Nigeria

Index

www.ingramcontent.com/pod-product-compliance
Lightning Source LLC
LaVergne TN
LVHW080625210926
848485LV00003B/212